T0215677

Pro Jakarta EE 10

Open Source Enterprise Java-based Cloud-native Applications Development

Peter Späth

Apress®

Pro Jakarta EE 10: Open Source Enterprise Java-based Cloud-native Applications Development

Peter Späth
Leipzig, Sachsen, Germany

ISBN-13 (pbk): 978-1-4842-8213-7   ISBN-13 (electronic): 978-1-4842-8214-4
https://doi.org/10.1007/978-1-4842-8214-4

Managing Director, Apress Media LLC: Welmoed Spahr
Acquisitions Editor: Steve Anglin
Development Editor: Laura Berendson
Coordinating Editor: Mark Powers
Copy Editor: Kezia Endsley

Cover designed by eStudioCalamar

Cover image by Dapiki Moto on Unsplash (www.unsplash.com)

Distributed to the book trade worldwide by Apress Media, LLC, 1 New York Plaza, New York, NY 10004, U.S.A. Phone 1-800-SPRINGER, fax (201) 348-4505, e-mail orders-ny@springer-sbm.com, or visit www.springeronline.com. Apress Media, LLC is a California LLC and the sole member (owner) is Springer Science + Business Media Finance Inc (SSBM Finance Inc). SSBM Finance Inc is a **Delaware** corporation.

For information on translations, please e-mail booktranslations@springernature.com; for reprint, paperback, or audio rights, please e-mail bookpermissions@springernature.com.

Apress titles may be purchased in bulk for academic, corporate, or promotional use. eBook versions and licenses are also available for most titles. For more information, reference our Print and eBook Bulk Sales web page at http://www.apress.com/bulk-sales.

Any source code or other supplementary material referenced by the author in this book is available to readers on GitHub (https://github.com/Apress). For more detailed information, please visit http://www.apress.com/source-code.

Printed on acid-free paper

To Nicole.

Table of Contents

About the Author

Peter Späth graduated in 2002 as a physicist and became an IT consultant, mainly for Java-related projects. In 2016, he decided to concentrate on writing books on various technology-related topics, with his main focus on software development. With two books about graphics and sound processing, two books about Android app development, and several books about Java, Späth continues his efforts in writing software development-related literature.

About the Technical Reviewer

Massimo Nardone has more than 25 years of experience in security, web/mobile development, cloud, and IT architecture. His true IT passions are security and Android. He has been programming and teaching others how to program with Android, Perl, PHP, Java, VB, Python, C/C++, and MySQL for more than 20 years. He holds a Master of Science degree in computing science from the University of Salerno, Italy. He has worked as a CISO, CSO, security executive, IoT executive, project manager, software engineer, research engineer, chief security architect, PCI/SCADA auditor, and senior lead IT security/cloud/SCADA architect for many years. His technical skills include security, Android, cloud, Java, MySQL, Drupal, Cobol, Perl, web and mobile development, MongoDB, D3, Joomla, Couchbase, C/C++, WebGL, Python, Pro Rails, Django CMS, Jekyll, Scratch, and more. He was a visiting lecturer and supervisor for exercises at the Networking Laboratory of the Helsinki University of Technology (Aalto University). He also holds four international patents (in the PKI, SIP, SAML, and Proxy areas). He is currently working for Cognizant as the head of cybersecurity and CISO to help clients in areas of information and cybersecurity, including strategy, planning, processes, policies, procedures, governance, awareness, and so forth. In June, 2017, he became a permanent member of the ISACA Finland Board. Massimo has reviewed more than 45 IT books for different publishing companies and is the co-author of *Pro Spring Security: Securing Spring Framework 5 and Boot 2-based Java Applications* (Apress, 2019), *Beginning EJB in Java EE 8* (Apress, 2018), *Pro JPA 2 in Java EE 8* (Apress, 2018), and *Pro Android Games* (Apress, 2015).

Introduction

Java is not just a programming language, it's also a platform used to host software. As far as enterprise environments go, Java Enterprise Edition Jakarta EE (formerly JEE) has an extensive collection of APIs that are particularly useful for addressing corporate IT requirements.

This book covers advanced topics of Jakarta EE development. This includes pro-level web tier development, architecture-related matters, advanced XML and JSON processing, application clients and scripting languages, resource handling, high-level security enhancements, and advanced monitoring and logging techniques.

The target version of Jakarta EE is version 10. Unless otherwise noted, server scripts have been tested on Ubuntu 22.04. Switching to Debian, Fedora, or OpenSUSE Leap should not impose any problems.

The book is for advanced enterprise software developers with knowledge of Java Standard Edition version 8 or later and some experience in Jakarta EE (or JEE) development. Having read the book titled *Beginning Jakarta EE* from the same author and publisher (ISBN: 978-1-4842-5078-5) will surely help, but it's not a strict prerequisite. I keep references to the beginner book at a bare minimum. I also assume that you can use the online API references, so this book is not a complete reference in the sense that not all API classes and methods are listed. Instead, the book includes techniques and technologies that will help professional Java enterprise-level developers deal with topics and master problems that arise in a corporate environment.

The book uses the Linux operating system as its development platform, although the code can run on other platforms without requiring complex changes. The server installations—like version control, continuous integration systems, and operating instructions—all target the Linux operating system. This book also does not cover hardware issues, except for some cases where hardware performance has a noticeable impact on the software.

After finishing this book, you will be able to develop and run Jakarta EE 10 programs of mid- to high-level complexity.

How to Read This Book

You can read this book sequentially from the beginning to the end, or you can read chapters on an ad hoc basis if your work demands special attention to one or another topic.

Source Code

All the source code in this book can be found at `github.com/Apress/pro-jakarta-ee10`.

PART I

Development Workflow Proposal

CHAPTER 1

Installing a Development Server

This book uses GlassFish version 7.0.1 as a Jakarta EE server, although I try to avoid vendor lock-in, so that, unless otherwise noted, you can test all the examples on different Jakarta EE servers.

For the Eclipse IDE, there is a GlassFish plugin called GlassFish Tools, which you can use if you want to. I don't use it in this book for several reasons. First of all, the plugin can have issues with your Eclipse installation. Second, if you don't use the plugin and instead use a terminal to start and stop the server and then use a build tool like Gradle to install and uninstall Enterprise applications, you are already close to what you need for integration testing and production setups. Third, it is easier to switch between different Jakarta EE servers and different IDEs. Fourth, you don't have to learn how to use that plugin, including any peculiarities that might arise.

So for now, simply download and install the GlassFish server, version 7.0.1, from this location:

```
https://glassfish.org/download.html
```

Select the Full Platform variant.

Note GlassFish 7.0.1 runs on JDK 11 through 19. You can try higher JDK versions for GlassFish 7.0.1, but they might not work.

© Peter Späth 2023
P. Späth, *Pro Jakarta EE 10*, https://doi.org/10.1007/978-1-4842-8214-4_1

Installing and Running Under Linux

If you have a Linux development box, unzip the GlassFish distribution to any folder suitable for your needs. I call this folder GLASSFISH-INST in this chapter.

GlassFish by default uses the default Java installed on your system. If you don't want that, open the GLASSFISH-INST/glassfish/config/asenv.conf file and add the following:

```
AS_JAVA=/path/to/your/sdk
```

to the end, replacing /path/to/your/sdk with your JDK installation directory.

In addition, make sure the JAVA_HOME environment variable points to the JDK installation directory:

```
# In any terminal for Glassfish operation, first run
export JAVA_HOME=/path/to/your/sdk
```

You can enter this line on top of the GLASSFISH-INST/bin/asadmin file, after the #!/bin/sh preamble. That way, you don't have to enter it each time you start a new terminal.

Even without any Jakarta EE server applications installed, it should be possible to start the server from a console. Enter the following:

```
cd GLASSFISH-INST
bin/asadmin start-domain
```

You should see the following as the command output:

```
Admin Port: 4848
Command start-domain executed successfully.
```

If you open a browser window at http://localhost:4848, you can see the web administrator frontend (the admin console); see Figure 1-1.

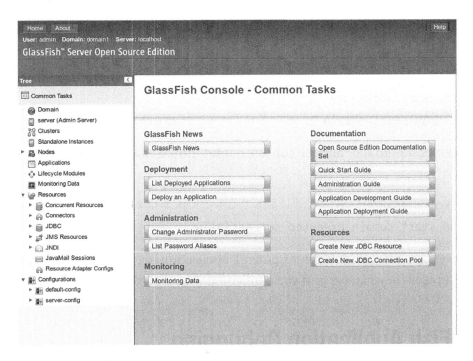

Figure 1-1. *GlassFish server running*

If you want to use a different port, you can change it in the admin console by choosing Configurations ➤ Server-config ➤ HTTP Service ➤ Http Listeners ➤ Admin-listener. You can also make this change via the terminal:

```
bin/asadmin set server-config.network-config.
network-listeners.network-listener.admin-listener.
port=4444
```

(Type this on one line when the server is running.) Then restart GlassFish and add the -p 4444 option to the asadmin invocations.

Installing and Running Under Windows

The GlassFish installation instructions for Windows are similar to those for Linux. Unzip the GlassFish distribution and add the following

```
set AS_JAVA=C:\path\to\your\sdk
```

to the end of the `GLASSFISH-INST/glassfish/config/asenv.bat` path. If you don't want to use the system's default Java version, set the `JAVA_HOME` environment variable accordingly. Then run this in the console:

```
chdir GLASSFISH-INST
bin\asadmin start-domain
```

The server should start and the console output should look similar to this:

```
...
Admin Port: 4848
Command start-domain executed successfully.
```

To access the web administrator frontend, open a browser window at `http://localhost:4848` (or whatever port you configured); see Figure 1-1.

GlassFish Application Debugging

In order to enable debugging for Jakarta EE applications in GlassFish, one possibility is to start the server using the `--debug` flag:

```
cd GLASSFISH-INST
bin/asadmin --debug start-domain
```

The default debugging port of GlassFish is `9009`.

You can also use the web administrator console at `htp://localhost:4848`. Choose Configurations ➤ Server-config ➤ JVM Settings ➤ General tab. Then check the Debug checkbox. In the Debug Options field, you can change the debugging options, like the port and others. You have to restart the server after you change any debugging settings.

Once debugging is enabled, you can use a debugger like the one included in the Eclipse IDE to set breakpoints and investigate the program flow step-by-step.

Elaborated GlassFish Operation Instructions

You will find verbose information about the `asadmin` command and other operation procedures in the online documentation at `https://glassfish.org/documentation`, as well as in the *Beginning Jakarta EE* book from the same author and publisher (ISBN: 978-1-4842-5078-5).

CHAPTER 2

Creating and Building Projects with Eclipse

Eclipse is an IDE (Integrated Development Environment) with a plethora of functionalities that help you develop Java Enterprise projects. It is freely available and you can use it for commercial and non-commercial projects free of charge.

Eclipse can be extended by plugins, many of which are developed by the community and are free to use. Plugins, however, might also come from vendors and you might have to buy licenses to use them. This book uses free plugins. If you feel tempted to try proprietary plugins, which under the circumstances might boost your development, visit the Eclipse marketplace at

```
https://marketplace.eclipse.org
```

Consult each plugin's documentation to learn more about such extensions.

Installing Eclipse

Eclipse comes in several variants. To download any of them, go to

```
https://www.eclipse.org/downloads/
```

This book uses the Eclipse IDE for Enterprise Java Developers variant.

Note If you choose to download the installer, you will be later asked for the variant. To select the Enterprise variant from the start, click the Download Packages link and choose the Enterprise version on the next page.

7

© Peter Späth 2023
P. Späth, *Pro Jakarta EE 10*, https://doi.org/10.1007/978-1-4842-8214-4_2

This book uses Eclipse version 2022-09, but you can use higher versions. Just remember that if you run into trouble without an obvious solution, downgrading to Eclipse 2022-09 is an option.

Use any installation folder suitable for your needs. Plugin installations and version upgrades go into that folder, so make sure you have appropriate file access rights. On my Linux box, I usually put Eclipse in a folder called:

```
/opt/eclipse-2022-09
```

(or whatever version you have). I make it writable to my Linux user, as follows:

```
cd /opt
USER=...  # enter user name here
GROUP=... # enter group name here
chown -R $USER.$GROUP eclipse-2022-09
```

This changes the ownership of all files of the Eclipse installation, and it makes sense for a one-user workstation. If instead you have different users for Eclipse, you can create a new group called eclipse and allow write access to that group:

```
cd /opt
groupadd eclipse
chgrp -R eclipse eclipse-2022-09
chmod -R g+w eclipse-2022-09
USER=...  # enter your username here
usermod -a -G eclipse $USER
```

The chgrp ... command changes the group ownership and the chmod ... command allows write access for all group members. The usermod ... command adds a particular user to the new group.

Note   You need to be root for these commands. Also note that the usermod command does not affect the currently active window manager session on the PC. That means you must restart your system or, depending on your distribution, log out and log in again for that command to take effect.

As a last step, you can provide a symbolic link to the Eclipse installation folder:

```
cd /opt
ln -s eclipse-2022-09 eclipse
```

This makes it easier to switch between different Eclipse versions on your system.

On a Windows system, the installer sets the access rights for you and it is usually possible for any normal user to install plugins. This, however, depends on the Windows version and your system's configuration. In a corporate environment, you can configure more fine-grained access rights using Window's own access rights management. This could include not allowing normal users to install plugins or upgrades and creating superusers for administrative purposes.

Configuring Eclipse

Upon startup, Eclipse uses the default Java version installed on your system. If it cannot find Java or if you have several Java versions installed, you can explicitly tell Eclipse which Java to choose. To this aim, open the ECLIPSE-INST/eclipse.ini file and add the following line:

```
-vm
/path/to/your/jdk/bin/java
```

Add this line directly above this -vmargs line:

```
...
openFile
--launcher.appendVmargs
-vm
/path/to/your/jdk/bin/java
-vmargs
...
```

The format of the eclipse.ini file depends on your Eclipse version. Visit https://wiki.eclipse.org/Eclipse.ini for the correct syntax. On that site, you will also find precise instructions for specifying the Java executable path. The syntax I show here is for Eclipse 2022-09.

On Windows PCs, you specify the path as used on Windows systems:

```
...
-vm
C:\path\to\your\jdk\bin\javaw
...
```

Don't use escaped backslashes (such as in C:\\path\\to\\...), as you would expect for Java-related files!

Adding Java Runtimes

Eclipse itself is a Java application, and in the preceding section, you learned how to tell Eclipse which Java to choose for its own interests. For the development, you have to tell Eclipse which Java to use for compiling and running applications.

To do so, note the paths of all JDK installations you want to use for Eclipse development. Then, inside Eclipse, choose Window ➤ Preferences ➤ Java ➤ Installed JREs. Eclipse is usually clever enough to automatically provide the JRE it used for its own startup. If this is enough for you, you don't have to do anything. Otherwise, click the Add button to register more JREs. In the subsequent wizard dialog box, select Standard VM as the JRE type.

Select the checkbox to mark your primary JRE and don't forget to click the Apply or Apply and Close button to register your changes.

Adding Plugins

Eclipse can be extended by many useful plugins. Some of them are necessary for your development, and some just help improve your development workflow. This book doesn't use too many extra plugins, and I provide plugin installation instructions when they are needed.

As an exception, you now learn how to install a plugin for Groovy language support. You will use Groovy as a scripting language for various purposes throughout the book. In your browser, navigate to

```
https://marketplace.eclipse.org/content/
    groovy-development-tools
```

(ignore the line break), find the Install badge, and drag and drop it on the Eclipse window (see Figure 2-1).

Figure 2-1. *Installation badge plugin*

As is usually the case for Eclipse plugins, you have to acknowledge the license terms and select the features. For the features, you can accept the default selection.

In previous versions of Eclipse, to use Gradle as a build tool, you needed a Gradle plugin. However, with Eclipse 2022-09, this plugin is included, so no additional actions are required here.

Note   In the beginner's book for Jakarta EE development, I used Maven as a build tool. It is not a contradiction to now switch to Gradle. Maven and Gradle use similar concepts and Gradle even knows how to import Maven projects. Also, it is a good idea to be familiar with both build frameworks.

Using Eclipse to Administer a Jakarta EE Server

Even without using a Jakarta EE server plugin like GlassFish Tools, it is of course desirable to be able to deploy applications on a running server from inside Eclipse.

In this section, you create a basic RESTful service, which simply outputs the date to see how this can be done. For this aim, inside Eclipse, choose File ➤ New ➤ Other and then choose Gradle ➤ Gradle project. Enter `RestDate` as the project name. In the Options pane, leave the Use Workspace Settings in effect and finish the wizard.

The project will then be generated, as shown in Figure 2-2.

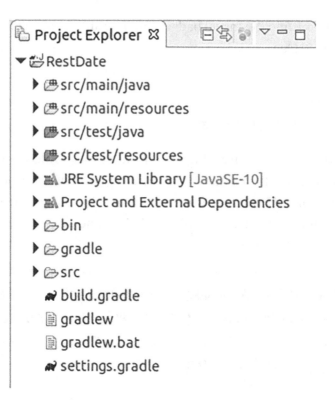

Figure 2-2. *Initial Gradle project*

Create a new package called book.jakartapro.restdate inside the src/main/java source section and create a class called RestDate. Enter the following code:

```
package book.jakartapro.restdate;

import java.time.ZonedDateTime;
import jakarta.ws.rs.GET;
import jakarta.ws.rs.Path;
import jakarta.ws.rs.Produces;

**
* REST Web Service
*/
@Path("/d")
public class RestDate {
  @GET
```

```
@Produces("application/json")
public String stdDate() {
  return "{\"date\":\"" +
      ZonedDateTime.now().toString() +
      "\"}";
}
}
```

Eclipse will complain about the imported classes because you haven't told it where to find them. As a remedy, open the build.gradle file and, inside the dependencies { } section, add the following:

```
implementation 'jakarta.platform:' +
    'jakarta.platform:jakarta.jakartaee-api:10.0.0'
```

Save the file and choose Gradle ➤ Refresh Cradle Project from the project's context menu. The error markers from Eclipse should now disappear.

You can explicitly tell Gradle to use the Java compiler version 1.17. To this aim, open build.gradle and add the following underneath the plugins { } section:

```
plugins {
  ...
}
sourceCompatibility = 1.17
targetCompatibility = 1.17
```

Choose Gradle ➤ Refresh Gradle Project in the project's context menu.

If you look at the project structure, you can see that a unit test section called src/test/java was also generated. For a simple REST controller like that, you can also create a unit test from the start to see it working correctly. Create a test class called book.jakartapro.restdate.TestRestDate inside the src/test/java section and let it read as follows:

```
package book.jakartapro.restdate;

import org.junit.Test;
import static org.junit.Assert.*;

public class TestRestDate {
    @Test public void testStdDate() {
```

```
    assertTrue("Unexpected date format",
        new RestDate().stdDate().matches(
            "\\{\"date\":\" +
            "\\d{4}-\\d{2}-\\d{2}T" +
            "\\d{2}:\\d{2}:\\d{2}\\..*"));
}
}
```

For any build task, the unit test will be executed by default. Gradle knows how to do that without further configuration, just because you placed the test classes in the `src/test/java` folder.

Now you need to let your RESTful application go to a WAR file, and for this aim, you need to add a deployment descriptor. You could avoid this by adding more annotations, but I personally like having a central configurations file, so that a web application's configuration is not splattered throughout the sources. Generate a folder called `src/main/webapp/WEB-INF` and add a file called `web.xml` to that folder:

```xml
<?xml version="1.0" encoding="UTF-8"?>
<web-app xmlns:xsi=
      "http://www.w3.org/2001/XMLSchema-instance"
   xmlns="http://xmlns.jcp.org/xml/ns/javaee"
   xsi:schemaLocation="http://xmlns.jcp.org/xml/ns/javaee
      http://xmlns.jcp.org/xml/ns/javaee/web-app_4_0.xsd"
   id="WebApp_ID" version="4.0">
  <display-name>RestDate</display-name>
  <servlet>
    <servlet-name>
      jakarta.ws.rs.core.Application
    </servlet-name>
  </servlet>
  <servlet-mapping>
    <servlet-name>
      jakarta.ws.rs.core.Application
    </servlet-name>
    <url-pattern>/webapi/*</url-pattern>
  </servlet-mapping>
</web-app>
```

14

In the same WEB-INF folder, add a beans.xml file. This is necessary to configure CDI (Context and Dependency Injection):

```xml
<?xml version="1.0" encoding="UTF-8"?>
<beans xmlns="https://jakarta.ee/xml/ns/jakartaee"
    xmlns:xsi="http://www.w3.org/2001/XMLSchema-instance"
    xsi:schemaLocation="https://jakarta.ee/xml/ns/jakartaee
        https://jakarta.ee/xml/ns/jakartaee/beans_3_0.xsd"
    bean-discovery-mode="all" version="3.0">
</beans>
```

You need one more deployment descriptor that's server product-specific. For GlassFish, you need a file called glassfish-web.xml with the following contents:

```xml
<?xml version="1.0" encoding="UTF-8"?>
<glassfish-web-app error-url="">
    <class-loader delegate="true"/>
</glassfish-web-app>
```

Note   If you are not deploying to GlassFish and instead are using a different Jakarta EE server, consult the server's documentation as to which server-specific deployment descriptors are needed.

The interesting part of the story begins now. You need to tell Gradle how to build a WAR from that and how to deploy it on a server. You first need to add a WAR plugin. Open the build.gradle file and enter id 'war' inside the plugins { } section:

```gradle
plugins {
    id 'war'
}
```

Choose Gradle ➤ Refresh Gradle Project from the project's context menu.

To build the WAR, open the Gradle Tasks view (choose Window ➤ Show View ➤ Other... to open it if you can't find it). Then right-mouse-click Build ➤ Build and select Run Gradle Tasks. The build task performs the unit tests and creates the WAR file. You can find the result in the build/libs folder.

Caution   Eclipse by default installs a filter to hide the Gradle build folder in the Project Explorer. To remove that filter, click the gray triangle tool in the header area of the Project Explorer, select Filters and Customization... and uncheck the Gradle Build Folder entry.

To deploy the WAR file on the server, you must first define some properties. Create the gradle.properties file in the project root and add the following, substituting your server's path:

```
glassfish.inst.dir = /PATH/TO/YOUR/GLASSFISH/SERVER
glassfish.user = admin
glassfish.passwd =
```

Also replace user and passwd according to your needs—change user to admin and use an empty password belong to a vanilla installation. If you use Windows, make sure you use double backslashes \\ instead of single ones for the path (since this is a Java properties file).

Next, add custom deploy and undeploy tasks to the build.gradle file. Open it and add the following:

```
task deployWar(dependsOn: war,
            description:">>> RESTDATE deploy task") {
  doLast {
    def FS = File.separator
    def glassfish =
        project.properties['glassfish.inst.dir']
    def user = project.properties['glassfish.user']
    def passwd = project.properties['glassfish.passwd']

    File temp = File.createTempFile("asadmin-passwd",
        ".tmp")
    temp << "AS_ADMIN_${user}=${passwd}\n"

    def sout = new StringBuilder()
    def serr = new StringBuilder()
    def libsDir = "${project.projectDir}${FS}build" +
        "${FS}libs"
```

```groovy
    def proc = """${glassfish}${FS}bin${FS}asadmin
        --user ${user} --passwordfile ${temp.absolutePath}
        deploy ${libsDir}/${project.name}.war""".execute()
    proc.waitForProcessOutput(sout, serr)
    println "out> ${sout}"
    if(serr.toString()) System.err.println(serr)

    temp.delete()
  }
}

task undeployWar(
            description:">>> RESTDATE undeploy task") {
  doLast {
    def FS = File.separator
    def glassfish =
        project.properties['glassfish.inst.dir']
    def user = project.properties['glassfish.user']
    def passwd = project.properties['glassfish.passwd']

    File temp = File.createTempFile("asadmin-passwd",
        ".tmp")
    temp << "AS_ADMIN_${user}=${passwd}\n"

    def sout = new StringBuilder()
    def serr = new StringBuilder()
    def proc = """${glassfish}${FS}bin${FS}asadmin
        --user ${user} --passwordfile ${temp.absolutePath}
        undeploy ${project.name}""".execute()
    proc.waitForProcessOutput(sout, serr)
    println "out> ${sout}"
    if(serr.toString()) System.err.println(serr)

    temp.delete()
  }
}
```

Both tasks specify a doLast { } block, which is a requirement for the tasks actually being executed in the task execution phase. If you didn't use such blocks, the corresponding code would be executed inside a configuration phase, regardless of whether the task is actually being executed or not. As you might guess, there is also a doFirst { } block possible. For the purposes here, you can use either of them.

The coding is for Linux as a development box. The change needed to run these same tasks under Windows is actually quite small. You only need to add a cmd /c to the beginning of the def proc = """... string:

```
// This is Windows!
def proc = """cmd /c ${glassfish}${FS} ...
```

If you want to use Windows, make the same change to both new tasks!

The code itself uses the Groovy language as an idiom. The project.properties addresses the properties defined in the gradle.properties file. The rest is Groovy code, with the .execute running a system shell command. The temporary file is needed, because for the asadmin command, it is only possible to hand over the password by specifying a specially tailored password file. The << writes into that file, yielding the necessary syntax as given by the asadmin specification.

In the Gradle Tasks view, go to the menu (inside the toolbar of the view) and check Show All Tasks. You can now run the deploy and undeploy tasks from that view. Navigate to Other and double-click either of the new custom tasks. The standard and error output is printed to the console view. Make sure the server is running before you deploy or undeploy the application.

Note   For any other Jakarta EE server, the deployment procedure will be very similar as seen from the build scripts. Because Groovy runs on top of Java, you can do a lot of other fancy things during deployment and undeployment. Besides, a task may easily call any other task and, by virtue of Gradle plugins, a lot of additional functionality is possible here.

To see this little application running, point your browser to http://localhost:8080/RestDate/webapi/d. The output will look like Figure 2-3.

JSON   Raw Data   Headers

Save Copy Collapse All Expand All ▽ Filter JSON

date: "2019-10-08T11:56:52.725+02:00[Europe/Berlin]"

Figure 2-3. *RESTful date application*

Note   To streamline development and save time deploying applications, you can also use the Gradle wrapper from a shell terminal. Go to the project's directory and then invoke `./gradlew deployWar` to perform a build and then a deploy.

Eclipse Everyday Usage

Eclipse provides a lot of functions and you can learn about them by opening the built-in help. Following is a list of tips that will help you to get most out of Eclipse:

- You can get to an identifier's definition by placing the cursor over it and pressing F3. This works for variables (navigate to its declaration) and classes/interfaces (navigate to its definition). You can even inspect referenced and Java standard library classes that way. Eclipse will possibly download sources and show the code. This is a great way to learn about libraries by looking at the code in-depth.

- To rapidly find a resource, such as a file, class, or interface, press CTRL+SHIFT+R.

- Start typing the code and then press CTRL+SPACE. Eclipse will show you suggestions for finishing your typing. For example, type new `SimpleDa` and then press CTRL+SPACE. The list provided will contain all the constructors for the `SimpleDateFormat` class. Even better, you can shortcut that by typing new `SiDF` and pressing CTRL+SPACE, because Eclipse will guess the missing lowercase letters. An additional goody is that you don't have to write the `import` statements for classes and interfaces you introduce that way—Eclipse will do the `imports` for you.

- Let Eclipse do the imports for all classes not yet resolved by pressing SHIFT+CTRL+O (think of O as "organize imports").

- Format your code by pressing CTRL+ALT+F. This also works for XML and other file types.

- Let Eclipse show you super- and subtypes by pressing F4 over a type designator.

- Use F5 to update the Project Explorer view, if files were added or removed from outside of Eclipse.

- With a new Eclipse installation, open the Problems view from Window ➤ Show View ➤ Other... ➤ General ➤ Problems. It will readily point you to any problems that Eclipse detects (compiler problems, configuration problems, and others).

- Open the Tasks view from Window ➤ Show View ➤ Other... ➤ General ➤ Tasks to get a list of all occurrences of TODO that you entered in code comments.

- If TODO is not enough fine-grained for you, you can add bookmarks by right-mouse-clicking anywhere on the vertical bar on the left side of the code editor. Bookmarks are listed in the Bookmarks view.

Changing the Project Layout

So far, you placed all file artifacts into the following:

```
Java sources -> src/main/java
Test sources -> src/test/java
Web files    -> src/main/resources
```

You didn't specify this in the build.gradle file, because these are the locations Gradle uses by default. By the way, the Maven build framework uses these same places. There are good reasons to deviate from this layout, though. The most prominent reason is you that you have a prescribed layout because a version control system checkout dictates it. It is not hard to tell Gradle where to look for files in such cases. In the build.gradle file, simply add the following:

```
sourceSets {
  main {
    java {
      srcDirs = ['path/to/folder']
    }
    resources {
      srcDirs = ['path/to/folder']
    }
  }
  test {
    java {
      srcDirs = ['path/to/folder']
    }
    resources {
      srcDirs = ['path/to/folder']
    }
  }
}
```

where you substitute the path/to/folder with your custom project layout folders. The [...] are Groovy language lists, so you can also place a comma-separated list there to designate several folders, which will then be merged.

CHAPTER 3

Development with NetBeans as an IDE

Most modern IDEs can fulfill your development needs, including Groovy and Gradle support, versioning, compiling and running Java, and more. I could include more "Using *XXX* as IDE" chapters and fill half of the book, but most IDEs follow similar concepts, have a help functionality, and provide tutorials and examples. The path to a development workflow usually follows these steps:

1. Learn how to create projects, files, and folders.

2. Learn how to edit files: Java, XML, resources, script files, and others.

3. Learn how to connect projects and JREs.

4. Learn how to add Gradle functionality to projects.

5. Learn how to add versioning.

6. Learn to use more, less important but helpful tools.

You learned how to use Eclipse, which provides all you need to develop and deploy Jakarta EE projects. As an example of a second IDE that you can use for development, this chapter briefly investigates another free IDE maintained by Apache, called NetBeans.

Installing NetBeans

Go to `https://netbeans.apache.org/` and download NetBeans. Choose one of the installers that suits your operating system. You can also install the binary version,

© Peter Späth 2023
P. Späth, *Pro Jakarta EE 10*, https://doi.org/10.1007/978-1-4842-8214-4_3

something like `netbeans-16-bin.zip`, but then you have to manually set some configuration settings that the installer otherwise does for you.

More installation options are described on the download page.

Note   Contrary to Eclipse, the NetBeans IDE stores some settings in a USER_ HOME/`.netbeans` folder. Keep that in mind if you want to uninstall NetBeans or reinstall it from scratch.

Starting a NetBeans Project

Choose File ➤ New Project... to generate a new project. Then select one of the Java with Gradle project templates:

- **Java Application**

  Generates a basic Java application template. If you use Gradle as a build and dependency resolution tool, you can start here and add features later.

- **Java Class Library**

  Use this to create a library JAR that can be used with other projects.

- **Web Application**

  This is a Java application plus an associated server. Since you'll continue controlling servers from the Gradle script in this book, this option is not explained. You can of course experiment with it and use it for your own purposes.

- **Multi-Project Build**

  This build creates a project with one or more sub-projects. This happens by using `include` directives inside `settings.gradle`. You can specify the sub-projects while installing the root project, and NetBeans will create one new project view per sub-project.

— **Java Frontend Application**

This creates a Java frontend application blueprint. For this installation option to work, NetBeans must be running under Java 11 or later (you can change this option in `NETBEANS-INST/netbeans/etc/net-beans.conf`).

— **Micronaut Project**

Generates a project based on the Micronaut platform (non-Jakarta EE).

Because it is enough for the purposes in this book, the following section uses the Java Application type.

Multi-IDE Projects

Chapter 2 used a Gradle project style. Because you'll do the same here for NetBeans, it is extremely easy to migrate a project from Eclipse to NetBeans and vice versa. Do you remember the simple `RestDate` project you created for Eclipse in Chapter 2? To move that project to NetBeans, you simply copy the `build.gradle`, `gradle.properties`, and `settings.gradle` files from `ECLIPSE-PROJ` (the project folder) to `NETBEANS-PROJ`, and then do likewise for all the `src` files. See Table 3-1.

Table 3-1. *Migrating Gradle Projects*

Eclipse		NetBeans
ECLIPSE-PROJ/	→	NETBEANS-PROJ/
build.gradle	→	build.gradle
settings.gradle	→	settings.gradle
gradle.properties	→	gradle.properties
src/*	→	src/*

First create a Java project by choosing Gradle ➤ Java Application in NetBeans. Check the Initialize Gradle Wrapper checkbox so the wizard will also generate a non-IDE (shell) script set in the project. Then perform the copy operations shown in Table 3-1.

Note   NetBeans detects the file changes and automatically downloads the new dependencies. This takes a minute or so, so give NetBeans some time to restore the project setup. Error markers won't disappear until everything is downloaded and reconfigured.

The new project view in NetBeans will now look similar to Figure 3-1. After you start the server, you can double-click the `deployWar` task to build and deploy the project.

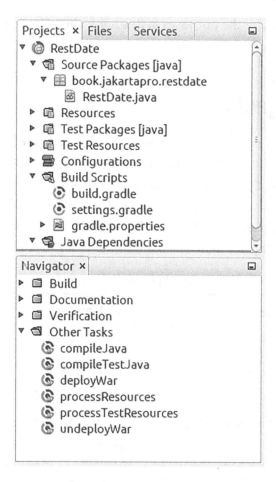

Figure 3-1. *RestDate migrated to NetBeans*

In the rest of this book, you use Eclipse, but in this chapter you have seen that migrating to and using a different IDE is easy.

CHAPTER 4

Git and Subversion

Version control systems serve several purposes. First, they are a kind of backup system. Once committed (or pushed, in the case of Git), a file has a persisted copy on some server, so if a developer's machine crashes, the files are not lost. Second, a version control system keeps a history of file versions, so an edit will not lead to previous versions being lost. Third, a continuous integration (CI) system might be connected to a version control system repository, which means development artifacts can be automatically published to integration systems for testing and then later released to some production (or pre-production) system.

This chapter covers the Subversion and Git version control systems. Other version control systems exist, but Git and Subversion are the most commonly used products.

Note For client-side usage only, skip to the "Subversion Clients" and "Git Clients" sections. Using Git and Subversion clients is well documented in manuals and tutorials. The rest of this chapter is about installing corporate Subversion or Git servers, which makes sense if you don't want to host sources outside your company.

The Subversion Version Control System

Subversion is a somewhat older version control system that's still actively maintained. It is a centralized version control system, which means that all versions are stored in a central place. Eclipse, and all other IDEs that I am aware of, provide built-in functionality to talk to Subversion or allow you to install plugins for that purpose.

Subversion is readily installed on a Linux server. To install it, run the following code as `root`:

© Peter Späth 2023
P. Späth, *Pro Jakarta EE 10*, https://doi.org/10.1007/978-1-4842-8214-4_4

```
# Ubuntu or Debian
apt-get install subversion

# Fedora or RHEL
yum install subversion

# SuSE
zypper install subversion
```

Unless the installation creates a subversion user and group, you can check this by running cat /etc/group | grep subversion and cat /etc/passwd | grep subversion. Then create a subversion user and group by entering the following:

```
# We are "root"

# Ubuntu, Debian, Fedora, RHEL
mkdir -p /srv/subversion
adduser --home /srv/subversion subversion
chown -R subversion.subversion /srv/subversion

# OpenSUSE Leap
mkdir -p /srv/subversion
useradd -r -U --home-dir /srv/subversion subversion
chown -R subversion.subversion /srv/subversion
```

As an example, you can create a repository for the RestDate project from Chapter 2:

```
# Create an empty repository
# We are "root"
su -l subversion
svnadmin create /srv/subversion/RestDate

# Creating the standard subversion layout
cd # going home
mkdir tmp
cd tmp
svn co file:///srv/subversion/RestDate
cd RestDate
mkdir trunk tags branches
```

```
svn add *
svn commit -m "Init"

# On Debian and OpenSUSE Leap:
chmod -R g+w /srv/subversion/RestDate

# Cleanup, we are going to work with a normal user
cd ..
rm -rf RestDate
exit
```

This creates a standard Subversion repository layout. The `trunk` is the main development branch, `branches` contains all the branches, and `tags` is for tags. This is the usual convention, but you are not obliged to follow it. It does help others understand your versioning procedures better if you follow this convention.

Subversion provides different methods for clients to connect to repositories. The following list shows a summary:

- `file:///`

  Local access only. Not really an option, because a corporate server should not be a developer's workstation at the same time.

- `http://`

  Indirect access via a mediating Apache HTTP server.

- `https://`

  Same as `http://`, but with additional SSL encryption.

- `svn://`

  Subversion's own network protocol.

- `svn+ssh://`

  Access mediated via an SSH login.

For several reasons, I highly favor the `svn+ssh://` protocol. First of all, it is straightforward and easy. There is no need for any other software to run on the server, and you don't need to spend time creating a bulletproof Apache installation, which is necessary for `http://` or `https://`, including maintenance. And you don't have to use the unsecured `svn://` protocol access. Security is delegated to the very safe SSH access method instead. The usual arguments against using SSH are not convincing—you just need a single SSH user for your Subversion work and don't waste disk space. Besides, you can use rbash to restrict shell access to this user. Because of SSH's importance, the community is highly motivated to immediately fix any SSH vulnerabilities that arise.

Of course, the decision is up to you and you will find installation instructions in the documentation and on the Internet. The rest of this chapter uses the svn+ssh:// protocol, because it is so easy.

You first need to make sure that logins using rbash as the shell in the subversion group can execute only a very limited set of commands. For this aim, you need to go through a few preparational steps, depending on the distribution you use:

```
# Fedora ********************************************
ln -s /bin/bash /bin/rbash

# OpenSUSE Leap ************************************
# Here PATH=/usr/lib/restricted/bin is the default path
# for executable commands for rbash. It cannot be
# changed. What you can do here is to further limit
# access to /usr/lib/restricted/bin based on group
# and ownership policies. For example
#     groupadd rbashusers
#     chgrp rbashusers /usr/lib/restricted/bin/*
#     chmod o-x /usr/lib/restricted/bin/*
# and NOT adding group "rbashusers" to user "svnuser"
# (see below). For this to work the links in that folder
# must be HARD links (no -s option in \ci{ln})
```

Next, add the svnuser.sh file inside /etc/profile.d/:

```
# Ubuntu, Debian or Fedora
if [ "$SHELL" = "/bin/rbash" -a \
    $(groups | grep -c \\bsubversion\\b) = 1  ]; then
  export PATH=.
fi

# On OpenSUSE Leap don't write this file. See comment in
# previous listing
```

This way, users with the rbash shell belonging to the subversion group, logging in via SSH, can only use a limited set of commands. Effectively, any Subversion user trying to log in to your server via ssh will soon realize that such a direct login makes no sense, because nothing can be done using the shell provided.

To create an SSH user for Subversion access, use the corresponding operating system functions and restrict the home directory to a bare minimum needed for SSH:

```
# We are "root"

# Ubuntu or Debian ***********************************
adduser --system --ingroup subversion \
    --shell /bin/rbash --disabled-password svnuser
ln -s /usr/bin/groups ~svnuser
ln -s /bin/grep ~svnuser
su svnuser --shell /bin/bash \
    -c "ssh-keygen -q -N '' -f /home/svnuser/.ssh/id_rsa"

# Fedora *********************************************
adduser -r -N -G subversion -s "/bin/rbash" \
    -d /home/svnuser svnuser
mkdir /home/svnuser
chown svnuser.subversion /home/svnuser
echo "export PATH=." > /home/svnuser/.bashrc
ln -s /bin/svnserve /home/svnuser
su svnuser --shell /bin/bash \
    -c "ssh-keygen -q -N '' -f /home/svnuser/.ssh/id_rsa"

# OpenSUSE Leap **************************************
useradd -r -G subversion -m -U -s /usr/bin/rbash svnuser
ln /usr/bin/svnserve /usr/lib/restricted/bin/svnserve
su svnuser --shell /bin/bash \
    -c "ssh-keygen -q -N '' -f /home/svnuser/.ssh/id_rsa"
```

The adduser creates the user, assigns the subversion group to the user, and, because of -shell /bin/rbash or -s, applies the access restrictions. The ssh-keygen builds the SSH key, and, what is more important for your purposes, it creates a space where you can put the public keys to connect from the outside.

For Ubuntu, Debian, and Fedora, I added a couple of command links in the home directory, because they are necessary for Subversion to work properly on these systems. For OpenSUSE Leap, the svnserve command is necessary, but I instead added a *hard* link of that svnserve command to the /usr/lib/restricted/bin folder.

Subversion Clients

As a client, you use the Subclipse Eclipse plugin. In your browser, navigate to

    https://marketplace.eclipse.org/content/subclipse

and drag the install badge onto your Eclipse window. Select all the features and finish the installation. Choose Window ➤ Preferences ➤ Team ➤ SVN and ignore the error message about a missing JavaHL installation. From the SVN Interface option, select SVNKit as the client. This will make the error message disappear. Click Apply and Close.

Now you need a public/private SSH key pair for each Subversion user on their workstation. If you already have one or if you have a Linux box, an ssh-keygen will create such a public/private key pair. On Windows, you can, for example, install Cygwin and use the same command, or use PuTTY (you must convert the key to the OpenSSH format in this case). It is also possible to install a key generator in the PowerShell. Use Install-Module -Force OpenSSHUtils and enter ssh-keygen.

The public counterpart of the key needs to go to the server. There, add its contents to the authorized_keys file inside the .ssh/ of the svnuser user you want to connect to. Make sure this file is visible to the user only:

```
# We are "root"
cat thePublicKey.file >>
    /home/svnuser/.ssh/authorized_keys
chown svnuser.subversion \
    /home/svnuser/.ssh/authorized_keys
chmod 600 \
    /home/svnuser/.ssh/authorized_keys
```

With the user's public key on the server, it is now possible to access the repository from Eclipse. To this aim, open the SVN Repository Exploring perspective by choosing Window ➤ Perspective ➤ Open Perspective ➤ Other... In the SVN Repositories tab on the left, right-mouse-click and choose New → Repository Location... For the URL, enter this:

```
svn+ssh://svnuser@THE.SERVER.ADDR/srv/subversion/RestDate
```

Unless Eclipse already knows about your SSH key or can guess it, you will be prompted for credentials and can point Eclipse to the key; see Figure 4-1.

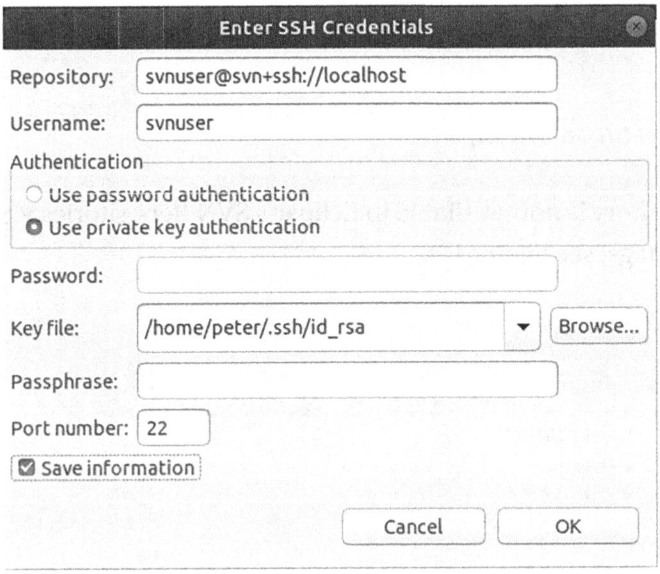

Figure 4-1. SSH credentials input

Caution   If you didn't add a passphrase to that key while building it, Eclipse shows another input dialog for that empty passphrase. Enter an x into the password field to enable the OK button. Then check the Save Password checkbox, clear the password field, and click OK. See Figure 4-2.

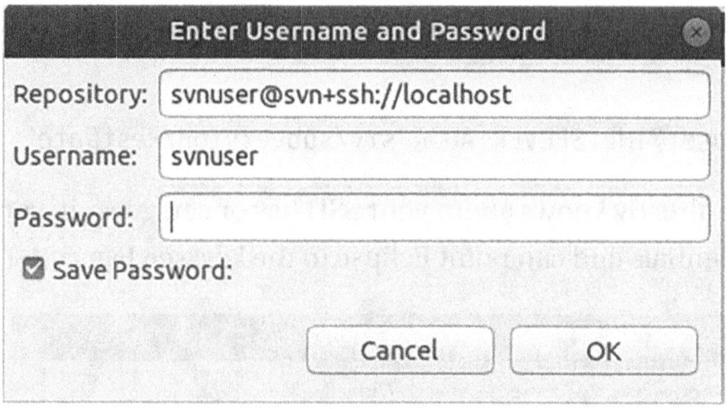

Enter Username and Password

Repository: svnuser@svn+ssh://localhost

Username: svnuser

Password: |

☑ Save Password:

Cancel   OK

Figure 4-2. *Extra password input*

The new repository is now available in Eclipse's SVN Repositories view, including all the branches and tags; see Figure 4-3.

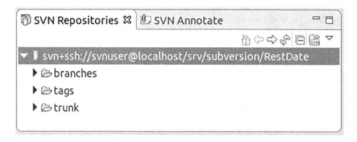

Figure 4-3. *Repository view*

To connect the RestDate project to Subversion, right-click the project, then choose Team ➤ Share Project... In the wizard dialog box, select SVN as the repository type, then check the Use Existing Repository Location option and select the repository you just registered. In the next wizard dialog box, check the Use Specified Folder Name option and enter trunk. See Figure 4-4. Click Next, then Finish to complete this process.

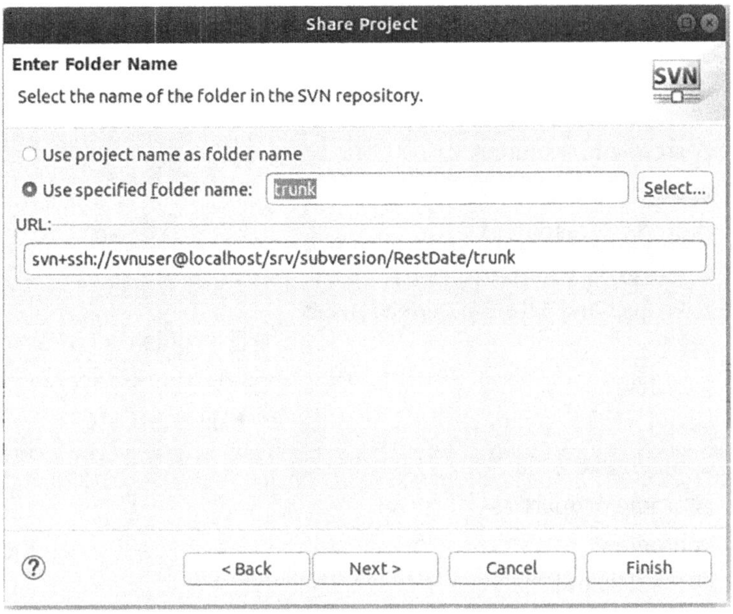

Figure 4-4. *Connect the project to SVN*

The project has now been shared with Subversion. You can see it by looking for [trunk], which was added to the Package Explorer view. See Figure 4-5.

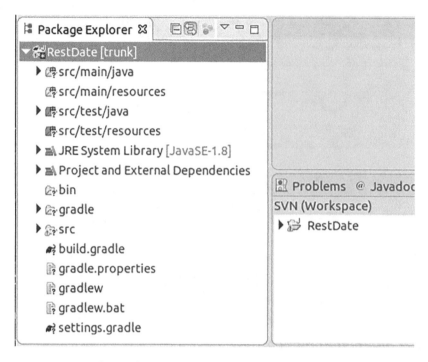

Figure 4-5. *Project under Subversion control*

Now you need to add all the files and folders—except for the `bin` folder—to the
Subversion repository. Right-mouse-click the `bin` folder and choose Team ➤ Add to
svn:ignore... Then select the project by right-mouse-clicking and choosing Team ➤
Commit... Then set all the other files and folders under version control.

The Git Version Control System

Git is a modern, distributed version control system. Distributed means you *might* have a
central repository, but any number of fully standalone repositories for the same project
can exist on different machines, including the developer's box.

This means that project teams situated in different locations with different network
access policies don't have to struggle with network differences, and development
and version control system operation is possible, even without network access. In a
distributed version control system like Git, it is therefore easier to run `local` repositories,
which only later can have their data moved to a central repository.

To install Git and a `git` user, a `git` group, and a base directory for repositories on a
Linux server, enter the following:

```
# Ubuntu or Debian
apt-get install git
mkdir -p /srv/git
adduser --home /srv/git git
chown -R git.git /srv/git

# Fedora or RHEL
yum install git
mkdir -p /srv/git
adduser --home /srv/git git
chown -R git.git /srv/git

# OpenSUSE Leap
zypper install git
mkdir -p /srv/git
useradd -U --home /srv/git git
chown -R git.git /srv/git
```

Next, you'll create an example repository. For demonstration purposes, you can use the RestDate project from Chapter 2:

```
# Create an empty repository
# We are "root"
su -l git
cd /srv/git
mkdir RestDate
cd RestDate
git init
exit
```

Git supports several connection protocols: file://... for local access (connecting to a repository on the same machine), http://... and https:// for connecting through a web server, git://... for using a protocol and port specially tailored for Git, and ssh://... for access mediated by an SSH account. While the git and http methods transfer unencrypted data and thus are out of scope, the file method is disqualified for remote access. That means you are left with https and ssh for commercial projects.

The https method requires a specially configured (Apache) HTTP server and leads to substantial extra effort and cost. It is used often with community-driven projects. In a corporate environment, using Git via SSH is the better option. It is heavily immune against intrusion, and you can configure a dedicated Git user with limited access to system resources:

```
# We are "root"

# Ubuntu or Debian **************************************
adduser --system --ingroup git \
    --shell /bin/rbash --disabled-password gituser
ln -s /usr/bin/groups ~gituser
ln -s /bin/grep ~gituser
su gituser --shell /bin/bash \
    -c "ssh-keygen -q -N ' ' -f /home/gituser/.ssh/id_rsa"

# Fedora ***************************************************
ln -s /bin/bash /bin/rbash
adduser -r -N -G git -s "/bin/rbash" \
    -d /home/gituser gituser
mkdir /home/gituser
chown gituser.git /home/gituser
echo "export PATH=." > /home/gituser/.bashrc
ln -s /usr/bin/git-upload-pack /home/svnuser
ln -s /usr/bin/groups /home/svnuser
ln -s /usr/bin/grep /home/svnuser
su gituser --shell /bin/bash \
    -c "ssh-keygen -q -N ' ' -f /home/gituser/.ssh/id_rsa"

# OpenSUSE Leap *************************************
useradd -r -G git -m -U -s /usr/bin/rbash gituser
ln /usr/bin/git-upload-pack \
    /usr/lib/restricted/bin/git-upload-pack
su gituser --shell /bin/bash \
    -c "ssh-keygen -q -N '' -f /home/gituser/.ssh/id_rsa"
```

This links the gituser account to the rbash shell, which means the gituser logging in from a terminal via an SSH shell has limited capabilities to use system resources. You can narrow that limit further by redefining the PATH variable. To this aim, create a file called gituser.sh inside /etc/profile.d:

```
# Ubuntu, Debian or Fedora
if [ "$SHELL" = "/bin/rbash" -a \
    $(groups | grep -c \\bgit\\b) = 1   ]; then
  export PATH=.
fi

# On OpenSUSE Leap don't write this file.
# PATH=/usr/lib/restricted/bin is the default path for
# executable commands for rbash on this system. It
# cannot be changed. What you can do here is to further
# limit access to /usr/lib/restricted/bin based on group
# and ownership policies. For example
#     groupadd rbashusers
#     chgrp rbashusers /usr/lib/restricted/bin/*
#     chmod o-x /usr/lib/restricted/bin/*
# and NOT adding group "rbashusers" to user "gituser"
# For this to work the links in that folder must be HARD
# links (no -s option in \ci{ln})
```

Users logging in via SSH cannot use commands from /bin, /usr/bin, or the other usual places for binaries and scripts, making the SSH shell effectively unusable.

Git Clients

The Git plugin, which the Eclipse IDE uses to talk to Git repositories, is included in the distribution, so you can use Git right from the start if you use Eclipse.

To connect to the Git repository on the server, you need a public/private SSH key pair for each Git user on their workstation. If you already have one or if you have a Linux box, ssh-keygen will create such a public/private key pair. On Windows, you can get one if you, for example, install Cygwin and use the same command, or use PuTTY (you must convert the key to the OpenSSH format in this case). There is also a module for PowerShell. Type Install-Module -Force OpenSSHUtils and enter ssh-keygen.

The public counterpart of the key needs to go to the server. There, add its contents to the authorized_keys file inside the .ssh/ of the gituser user you want to connect to. Make sure this file is visible to the user only:

```
# We are "root"
cat thePublicKey.file >>
    /home/gituser/.ssh/authorized_keys
chown gituser.git \
    /home/gituser/.ssh/authorized_keys
chmod 600 \
    /home/gituser/.ssh/authorized_keys
```

You now create a local Git repository as a clone of the server repository. Because of the distributed nature of Git, the relation to a server repository is much looser compared to centralized version control systems like Subversion. You could even start from a local repository, totally ignorant of the server repository, and only later make the connection. But with the server repository at hand, you can start with the cloning process.

To this aim, open the Git perspective by choosing Window ➤ Perspective ➤ Open Perspective ➤ Other... In the Git Repositories tab on the left, select the Clone a Git Repository link. In the wizard dialog box that appears, enter the following as the URI:

```
ssh://gituser@the.serv.er.addr/srv/git/RestDate
```

Substitute the.serv.er.addr with the Git server's address. In the authentication area, leave gituser as the user and don't enter a password. It is not needed, since you use the SSH key. See Figure 4-6.

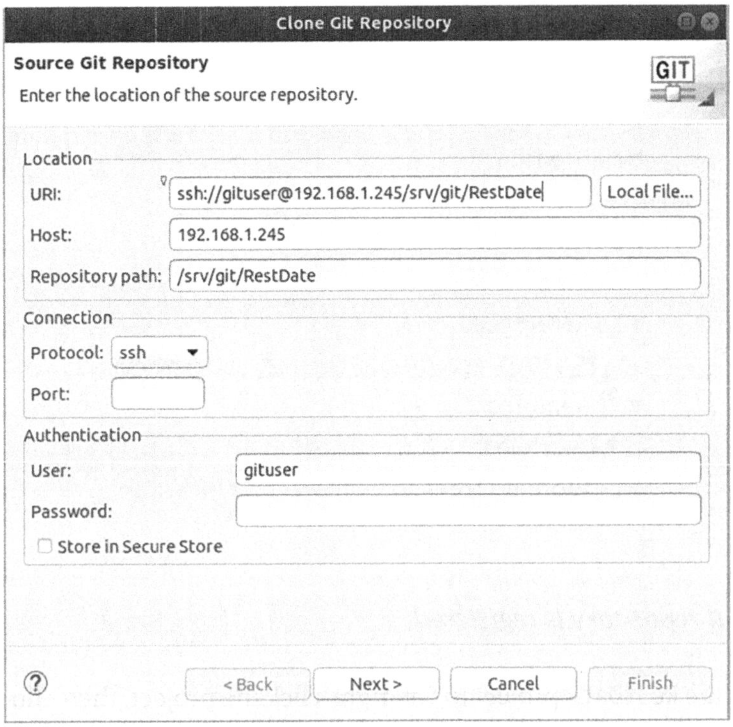

Figure 4-6. *Git repository input*

Click Next >, ignore the warning that the repository is empty, and click Next > again. In the last wizard dialog box, select a directory where you'll store the local Git repository. The usual convention is to *not* choose a directory inside the Eclipse workspace, to avoid messing up the workspace with Git meta information files. Click Finish.

The new repository now shows up in the repositories view. See Figure 4-7.

Figure 4-7. *Git repository is registered*

To connect the RestDate project to Git, right-click the project, then choose Team ➤ Share Project... In the wizard dialog box, select Git as the repository type, then select the local repository from the dropdown box. Then click Finish.

The project is now backed up by the local repository, and you can operate Git by right-mouse-clicking (on project) and choosing Team.

## CHAPTER 5

Continuous Integration

Continuous Integration (CI) involves continuously delivering applications to an integration server. A CI product on that server will then instantaneously or at least periodically deploy or redeploy such applications for testing and demonstration purposes.

This book doesn't talk about CI servers too much, since CI is not part of the Jakarta EE specification. Most CI products hook into version repository commits, however; so from a developer's perspective, CI is automatically performed once the developer performs a commit into Subversion or Git. Corporate policy usually determines how artifacts need to be checked in, more precisely into which branch they need to be checked in and how commits are marked, in order for CI to happen.

The Jenkins CI Server

Without going into detail, I present with a working example of a Jenkins installation. Jenkins is a free, open-source CI server. You can get the program and the user handbook from

```
https://jenkins.io/download/
https://jenkins.io/doc/book/
```

The rest of this chapter assumes that you are running a Jenkins server. The handbook tells you more in its "Installing Jenkins" chapter.

Caution Jenkins runs on port 8080. Since this port is frequently used by other programs as well, you can also configure a different port. Open the `/etc/default/jenkins` file and change the port at HTTP_PORT=8080 to a port that better suits your needs. Run `systemctl restart jenkins` to apply the changes. If Jenkins didn't start because port 8080 was used by a different

© Peter Späth 2023
P. Späth, *Pro Jakarta EE 10*, https://doi.org/10.1007/978-1-4842-8214-4_5

program, change the configuration and then issue a `systemctl start jenkins` command.

To see whether Jenkins has started correctly, you can view its log file at `/var/log/jenkins/jenkins.log`. In your browser, open `http://localhost:8080` and follow the instructions.

Note   If you see a blank page, run `systemctl restart jenkins` once to clean things up.

You will see a localized version of the Jenkins site. To get as close as possible to the descriptions in the following paragraphs, you can change the language setting in your browser to English. In Firefox, for example, enter `about:config` in the Address field, then change the `intl:accept_languages` setting. Put en in front of the comma-separated list.

Starting a Jenkins Project

Once you enter the Jenkins web administrator, create a new item using the New Item menu entry. This corresponds to a Jenkins project. Give it any name you like and choose Pipeline for the type. All the other types can be used for CI projects as well, but the pipeline type provides a good mixture of ease and flexibility. For the other types, consult the documentation.

Note   If you use the Jenkins WAR file installation type, you might have to add the plugins before you can proceed. This chapter assumes that you are using the CLI version.

As an example, you can use the `RestDate` project you created in Chapter 2 and create a corresponding Git repository directly on the Jenkins server:

```
sudo su        # become root
mkdir /srv     # only if it does not exist yet
mkdir /srv/git # only if it does not exist yet
mkdir /srv/git/RestDate
```

```
cd /srv/git
adduser git   # if missing yet
chown git.git RestDate
cd RestDate
su git           # now we are the git user
git init --bare
mkdir tmp        # temporary
cd tmp
git clone file:///srv/git/RestDate
cd RestDate
```

This script creates the folder, registers a new git user in the operating system, and initializes the Git repository. This assumes that adduser also creates a group of the same name. Ubuntu does that. Otherwise, use addgroup git to create that group. In the tmp folder, create a local clone of the repository. Now copy the gradle.properties, build. gradle, settings.gradle, gradlew, src, and gradle files and folders from the RestDate project to the tmp/RestDate folder.

Note   The gradlew file and the gradle folder exist only if you allowed Eclipse to build the Gradle wrapper. Make sure you did not uncheck the corresponding checkbox while creating the Eclipse project.

Add another file called Jenkinsfile (no suffix) with the following contents to the same folder:

```
pipeline {
    agent any
    stages {
        stage('Stage 1') {
            steps {
                echo 'Hello world!'
            }
        }
    }
}
```

This file describes a Jenkins pipeline. In the terminal, continue with the following:

```
git add *
git commit -m "Initial"
git push origin
```

This registers all the project files in the version control system (in the master branch, to be precise).

Back in the Pipeline configuration section of the Jenkins web administrator, change the definition to Pipeline script from SCM and specify the following:

```
SCM:                GIT
Repository URL:     file:///srv/git/RestDate
Credentials:        -none-
Branches to build:  */master
Repository browser: (Auto)
Script path:        Jenkinsfile
```

Obviously, if you had not created the Git repository on the same machine, you could have specified an appropriate Git URL such as ssh://..., together with an authentication mechanism, such as username plus a password or a certificate.

Build Triggers

So how does Jenkins know when to perform the next build and deployment? This is what the Build Triggers configuration section is for, as shown in Figure 5-1.

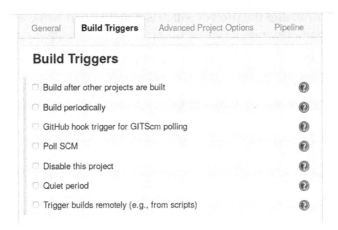

Figure 5-1. *Build Triggers section*

An elegant trigger option is to use the Trigger Builds Remotely method. Once you check it, you learn which URL to choose for the trigger. For this project and branch, this will be the following:

`http://my.jenkins.server:8080/job/My%20Project/build?token=TRGG5474Fhee5h`

where `TRGG5474Fhee5h` is a token added to the configuration side. See Figure 5-2.

Figure 5-2. *Remote build trigger*

Click the Save button and the project will be set up to do its work, although because of the very simple "Hello World" style Jenkinsfile, the build will not do anything really interesting. Nevertheless, you can click the Build Now link on the left-side menu to see what happens. The build you configured will be performed and the Stage View or the Full Stage View dialog will show a graphical timeline that includes the build. See Figure 5-3.

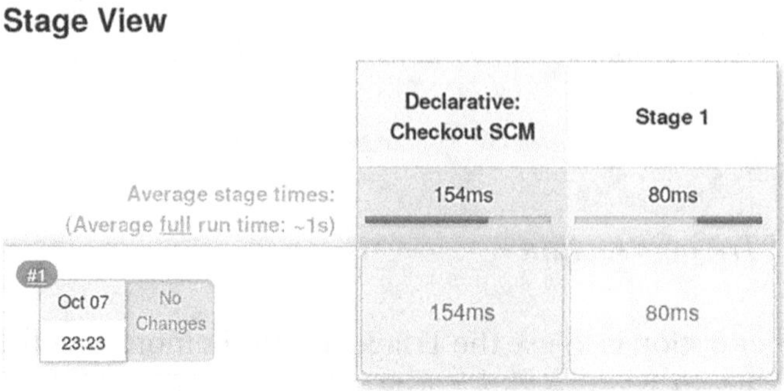

Stage View

Figure 5-3. *The build timeline*

This build has two parts—a Checkout SCM part and a Stage 1 part. The first belongs to a checkout of data from the version control system; the second was configured in the Jenkinsfile file. If you hover the mouse over the lower green part of the box representing a build part, a link appears that leads to the logs for this part. The logs for the source code checkout may read as follows:

```
No credentials specified
Cloning the remote Git repository
Cloning repository file:///srv/git/RestDate
 > git init /var/lib/jenkins/workspace/My Project
Fetching upstream changes from file:///srv/git/RestDate
 > git --version
 > git fetch --tags --progress --
   file:///srv/git/RestDate
   +refs/heads/*:refs/remotes/origin/*
 > git config remote.origin.url file:///srv/git/RestDate
 > git config --add remote.origin.fetch
   +refs/heads/*:refs/remotes/origin/*
 > git config remote.origin.url file:///srv/git/RestDate
```

```
Fetching upstream changes from file:///srv/git/RestDate
 > git fetch --tags --progress --
   file:///srv/git/RestDate
   +refs/heads/*:refs/remotes/origin/*
 > git rev-parse refs/remotes/origin/master^{commit}
 > git rev-parse refs/remotes/origin/origin/master^{commit}
Checking out Revision c71e0... (refs/remotes/origin/master)
 > git config core.sparsecheckout
 > git checkout -f c71e0...
Commit message: "X"
First time build. Skipping changelog.
```

You can see that Jenkins fetched the source code from the Git repository and saved the sources in the /var/lib/jenkins hierarchy. The logs from the other part (Stage 1) read as follows:

```
Hello world!
```

They exactly correspond to echo 'Hello world!' from the Jenkinsfile.

Creating Real-World Builds

In this section, you revert to the Jenkinsfile and add a real-world scenario to the build:

- A build stage, including unit tests

- A deployment stage

- An integration test stage

- A release stage

For this to work, add the following contents to the /srv/git/RestDate/tmp/ RestDateJenkinsfile file. (Remember that the Git checkout was placed in /srv/git/ RestDate/tmp/RestDate.)

```
pipeline {
    agent any
    stages {
        stage('Build Stage') {
```

```
            steps {
                echo 'Build Stage'
            }
        }
        stage('Deployment Stage') {
            steps {
                echo 'Deployment Stage'
            }
        }
        stage('Integration Test Stage') {
            steps {
                echo 'Integration Test Stage'
            }
        }
        stage('Release Stage') {
            steps {
                echo 'Release Stage'
            }
        }
    }
}
```

First implement the build stage. The Jenkinsfile allows an sh '...' directive to execute any shell script. You just add gradle WAR build and write the following:

```
...
stage('Build Stage') {
    steps {
        echo 'Build Stage'
        sh './gradlew war'
    }
}
...
```

The deployment stage assumes you have a GlassFish installation in /srv/ glassfish7. It also assumes that the GlassFish server is running and that the admin user doesn't have a password. You can then write the following:

```
...
stage('Deployment Stage') {
    steps {
        echo 'Deployment Stage'
        sh '/srv/glassfish6/bin/asadmin \
            undeploy RestDate || exit 0'
        sh '/srv/glassfish6/bin/asadmin \
            deploy build/libs/RestDate.war'
    }
}
...
```

The `...  || exit 0` code ensures that an undeploy of a not-deployed `RestDate` won't cause the stage to fail.

Note   The deployment succeeds only if the `jenkins` user (Jenkins runs under this user) is allowed to operate the `asadmin` command.

For an integration test, call the REST interface and check whether its output is a valid date and time string:

```
...
stage('Integration Test Stage') {
    steps {
        echo 'Integration Test Stage X'
        script {
            def result = sh(script: 'curl -X GET \
                "http://localhost:8080/RestDate/webapi/d"',
                returnStdout: true)
            def p = /\d{4}-\d{2}-\d{2}T\d{2}:\d{2}:\d{2}/
            assert result.trim() ==~
                /\{"date":"/ + p + /.*"\}/
        }
    }
}
...
```

This assumes that cURL is installed on your system. If it is missing, run this as root on Ubuntu or Debian: `apt-get install curl` (run `yum install curl` on Fedora or RHEL, and run `yast -i curl` on SuSE). Everything inside `script { ... }` is a Groovy script. The `sh ( ... )` executes a shell script. The `returnStdout: true` makes sure the output, not the status, will be returned from the `sh -call`. The `assert` is a Groovy language construct; it will use a regular expression to check whether the output is a date/time string.

Caution   The Groovy included with Jenkins is not fully fledged Groovy; some security restrictions apply. For this reason, you cannot use Groovy directly to perform network access. Instead, you need to revert to calling a shell function (the `sh( ... )`).

The last stage, called the release stage, does whatever is necessary to move the WAR to a release server. If you want to install the WAR file on a server using `scp`, for example, the definition would read as follows:

```
...
stage('Release Stage') {
    steps {
        echo 'Release Stage'
        sh 'scp build/libs/RestDate.war \
            user@the.server.addr:/path/to/RestDate.war'
    }
}
...
```

You must replace `user@the.server.addr` with the username and the server you want to copy the file to. Likewise, `/path/to/RestDate.war` is the location at the target server. For this to work, you must allow the user to use `scp` without password entry; for example, by moving the `jenkins` user's SSH key to the target server:

```
su jenkins
cd
ssh-keygen
# Copy the contents of .ssh/id_rsa.pub and paste it
# to the end of .ssh/authorized_keys on the target
# server
```

Caution   The .ssh/authorized_keys file must be restricted to owner-access only; otherwise, a standard SSH won't use it. Enter chmod 600 authorized_ keys to do that.

With all those changes in the Jenkinsfile file, commit and push the file to Git via the following:

```
git add Jenkinsfile
git commit -m "New Jenkinsfile"
git push origin
```

Back in the Jenkins web administrator, click the Build Now menu link from the project view. In the Stage View area, you should now see something like Figure 5-4.

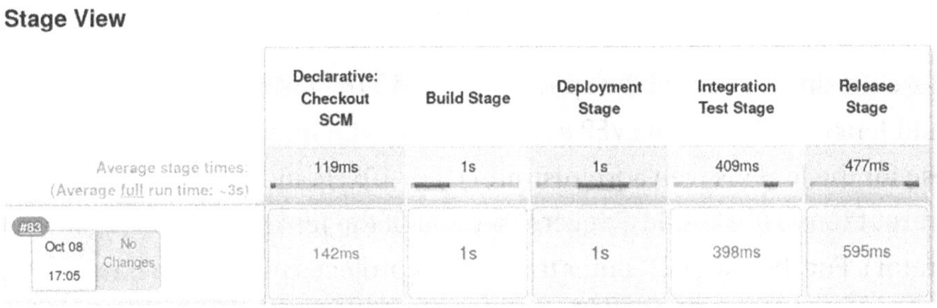

Stage View

	Declarative: Checkout SCM	Build Stage	Deployment Stage	Integration Test Stage	Release Stage
Average stage times: (Average full run time: ~3s)	119ms	1s	1s	409ms	477ms
#83 Oct 08 17:05 No Changes	142ms	1s	1s	398ms	595ms

Figure 5-4. *A successful full build*

From here, you can also view all logs from all stages. Hover the mouse over the green boxes and select the Logs link.

Triggering Builds from Git

In the previous "Build Triggers" section of this chapter, you learned how to configure an URL endpoint where you can trigger Jenkins builds from the outside. One possibility is to tell Git to invoke this trigger every time a new version is pushed into the central Git repository. To this aim, open this file:

```
/srv/git/RestDate/hooks/post-receive
```

If that file doesn't exist, create it and make it executable: `chmod a+x post-receive`. For its contents, write the following:

```
#!/bin/bash

USERPW=USER:PASSWD
SERVER=localhost
PORT=8712
TOKEN=TRGG5474Fhee5h
PROJECT=My%20Project

date >> jenkins-triggers.log
curl -X GET -u ${USERPW} -s -S \
  http://${SERVER}:${PORT}/job/${PROJECT}/build?
  token=${TOKEN} >> jenkins-triggers.log 2>&1
echo "Hook post-receive performed (informed CI)"
```

(Remove the line break and the spaces after `build?`.) For USER and PASSWD, you must write a valid Jenkins user. For SERVER and PORT, write the Jenkins server and port (you are using these for the Jenkins web administrator). For TOKEN, include the string you entered at Authentication in the "Build Triggers" section of the Jenkins project (again in the web administrator). For the PROJECT, enter the Jenkins project's name (you have to escape spaces as %20 shown here). The -s -S disables cURL's progress meter, but it allows error messages to be written. The echo message goes to the Git client, so committers see it after they push something to the central repository.

The trigger is ready now. Try it and push any change to the Git repo. The trigger will fire the Jenkins project and the web administrator will automatically show the new build process that was triggered. Make sure you have the project view open to see these activities immediately.

There are many more hooks you can use in Git to do interesting things. You can for example invoke some tests before the push actually happens, send the committer more information by -mail, and lots more. The possibilities are endless here.

Triggering Builds from Subversion

Once the remote triggers are activated for the Jenkins project, you can let Subversion trigger Jenkins builds just as you did for Git. Go to this path:

```
/path/to/svn/repo/hooks
```

Replace /path/to/svn/repo with the location of the Subversion repository. Create a new file called post-commit and write as its contents:

```
#!/bin/bash

USERPW=USER:PASSWD
SERVER=localhost
PORT=8712
TOKEN=TRGG5474Fhee5h
PROJECT=My%20Project
WHEREAMI=/path/to/the/repo

export PATH=/usr/local/sbin:/usr/local/bin
export PATH=${PATH}:/usr/sbin:/usr/bin:/sbin:/bin
cd $WHEREAMI

date >> jenkins-triggers.log
curl -X GET -u ${USERPW} -s -S \
  http://${SERVER}:${PORT}/job/${PROJECT}/build?
  token=${TOKEN} >> jenkins-triggers.log 2>&1
echo "Hook post-receive performed (informed CI)" \
  >> jenkins-triggers.log
```

(Remove the line break and the spaces after build?.) For USER and PASSWD, you must write a valid Jenkins user. For SERVER and PORT, provide the Jenkins server and port (you are using these for the Jenkins web administrator). For the TOKEN, include the string you entered at Authentication in the "Build Triggers" section of the Jenkins project (again in the web administrator). For PROJECT, include the Jenkins project's name (you have to escape spaces, as %20 here). For WHEREAMI, provide the location of the script. The -s -S disables cURL's progress meter, but allows error messages to be written.

Don't forget to make this file executable by entering `chmod a+x post-commit`. The current directory and the `PATH` environment variable had to be set because Subversion doesn't provide this information to the script.

In the Jenkins web administrator, update the project configuration and, in the Pipeline section, change the `Jenkinsfile` (pipeline) definition source to Pipeline Script from SCM and change SCM" to Subversion. As the repository URL, enter the `svn+ssh://USER@SERVER//path/to/repo/RestDate/trunk` string, where `USER` needs to be replaced with a username able to connect via SSH, `SERVER` is the server address, and `/path/to/repo` is the repository path. As in the previous section for Git, the `RestDate/trunk` is supposed to contain the complete Gradle project, including a `Jenkinsfile` file.

In the Credentials field, add the login information for using SSH. The possibilities include username plus password, or a username (probably `Jenkins`) plus a private SSH key (you must explicitly provide a key—take it from `jenkins/.ssh/id_rsa`). In the latter case, you must add the contents of `jenkins/.ssh/id_rsa.pub` to the `.ssh/authorized_keys` file on the server where the Subversion repository resides.

Caution   The `.ssh/authorized_keys` file must be restricted to owner-access only; otherwise, a standard-SSH won't use it. Enter `chmod 600 authorized_keys` to do that.

The trigger is now finished. Try committing an arbitrary change. Jenkins should start the project's build pipeline.

Analyzing Jenkins Builds via REST

You don't have to use the web administrator to monitor Jenkins builds. Instead you can use a REST API provided by Jenkins. This is extremely powerful for enterprise applications where you need scripted access to Jenkins.

To try that, create a Groovy project inside Eclipse, add a folder called `scripts` to the project, and announce that `scripts` folder to Eclipse by right-mouse-clicking the project and choosing Groovy Compiler. Check the Enable Project Specific Settings. Then click Add, enter `scripts/**`, and click Apply.

Create a `scripts/builds.groovy` file and enter the following script:

```
@Grapes(
```

```groovy
  @Grab(group='org.codehaus.groovy',
    module='groovy-all', version='2.5.8', type='pom')
)

import groovy.json.*

SERVER="localhost"
PORT=8080
USER="TheUser"
PASSWD="ThePassword"
PROJECT="My%20Project"

/**
 * Performs a GET given some URL. If the parameter does
 * not start with "http://", a "http://server:port/" gets
 * prepended. The requested content type is
 * 'application/json'
 */
def fetch(def url) {
  def enc64 = { s ->
    new String(Base64.encoder.encode(s.bytes)) }
  def c = ( !url.startsWith("http://")?
    new URL("http://${SERVER}:${PORT}/${url}"):
    new URL(url) ).openConnection()
  c.with {
    setRequestProperty('User-Agent','groovy')
    setRequestProperty('Accept','application/json')
    setRequestProperty('Authorization',
        "Basic " + enc64("${USER}:${PASSWD}"))
  }
  [output:c.inputStream.text, respCode:c.responseCode]
}

def d = fetch "job/${PROJECT}/api/json"
def json = new JsonSlurper().parseText(d.output)

// 'json' is the REST call result parsed into a
```

```
// Groovy object. See the JsonParser documentation in the
// Groovy docs
def lastTen = json.builds.sort{-it.number}.take(10)
//println lastTen
lastTen.each { b ->
  def url = b.url
  def build = new JsonSlurper().parseText(
        fetch(url + "api/json").output)
  //println build
  def author = build.changeSets ?
    build.changeSets[0]?.items[0]?.author?.fullName :
    "<undef>"
  println("" + build.number +
    " - " + new Date(build.timestamp) +
    " - " + build.result +
    " - " + author)
}
```

Replace the connection information at top of the file. This script fetches the last ten builds using the job/${PROJECT}/api/json API call, and then for each build, it issues another API call (the URLs are derived from the first call's results). It takes the build number, the build date, the outcome, and if the build comes from changed sources, the (first) author of the changes.

Note   The GRAPE library fetch on top of the script is necessary, because the Groovy plugin in Eclipse corresponds to the "Indy" variant of Groovy, missing some APIs. Make sure the version specified here matches the plugin's Groovy version (look in the preferences).

The REST API of Jenkins is much bigger compared to what you used in this section. You can get more information if you click the REST API link found at the bottom of most pages of the web administrator. The output of your script might look like this:

```
90 - Wed Oct 09 13:38:12 CEST 2019 - SUCCESS - Mark
89 - Wed Oct 09 13:09:02 CEST 2019 - SUCCESS - <undef>
```

```
88 - Wed Oct 09 13:06:52 CEST 2019 - FAILURE - <undef>
87 - Wed Oct 09 13:05:17 CEST 2019 - FAILURE - <undef>
86 - Wed Oct 09 10:06:12 CEST 2019 - SUCCESS - <undef>
85 - Wed Oct 09 09:52:02 CEST 2019 - SUCCESS - Pete
84 - Wed Oct 09 09:49:42 CEST 2019 - SUCCESS - Mark
83 - Tue Oct 08 17:05:04 CEST 2019 - SUCCESS - <undef>
82 - Tue Oct 08 17:01:58 CEST 2019 - FAILURE - Mark
81 - Tue Oct 08 16:02:42 CEST 2019 - SUCCESS - Mark
```

You don't have to use a fully-fledged scripting engine like Groovy to access the REST API, though. To access the API, you can also use cURL from a bash shell. You could start with the following sample request:

```
curl -X GET \
    -H "Accept: application/json" \
    -u USERNAME:PASSWORD \
    http://SERVER:PORT/job/PROJECT/api/json
```

where you must substitute SERVER, PORT, PROJECT, USERNAME, and PASSWORD with the appropriate values (don't forget to replace the spaces in the project name with the %20 escape sequence).

CHAPTER 6

Corporate Maven Repositories

Maven provides a repository and a dependency resolution build system for projects. You use it for Java and build related configuration files, and since Gradle uses Maven repositories, it is worthwhile to let some light shed on it to further streamline your Java Enterprise development process.

In the dependencies { ... } section of the build.gradle file, this script determines which libraries are needed for a project to be tested, built, and run:

```
...
dependencies {
    // This dependency is exported to consumers, that
    // is to say found on their compile classpath.
    api 'org.apache.commons:commons-math3:3.6.1'

    // This dependency is used internally, and not
    // exposed to consumers on their own compile
    // classpath.
    implementation 'com.codepoetics:protonpack:1.16'

    // This dependency is only needed for compilation.
    // It will not be packed in assemblies like WAR
    // files. It will also not be forwarded as a
    // dependency when publishing
    compileOnly 'jakarta.jakartaee-api:10.0.0'

    // Use JUnit test framework
    testImplementation 'junit:junit:4.13.2'
}
...
```

© Peter Späth 2023
P. Späth, *Pro Jakarta EE 10*, https://doi.org/10.1007/978-1-4842-8214-4_6

Furthermore, in the `repositories { ... }` section of the same file, you can see where to load the dependencies:

```
...
repositories {
    jcenter()
}
...
```

You can also declare one of `mavenCentral()` or `google` here. They, by convention, resolve to the following:

```
mavenCentral -> https://repo.maven.apache.org/maven2/
jcenter      -> https://jcenter.bintray.com/
google       -> https://maven.google.com/
```

The chapter title is "Corporate Maven Repositories," which means Maven has repository available only from a corporate intranet. What good would that be? First of all, if you are developing software for a space rocket or a nuclear power plant, you probably don't want to allow direct access to the public repositories. Second, if your company built its own libraries that are not supposed to be publicly available, you don't want to use a public repository either. So you'll at least want to think about using corporate Maven repositories.

If your company has decided to install a corporate Maven repository, you can use one of the products available for that purpose. Right? Maybe not. From a client perspective, a repository is just a web server that provides static content (the JARs and some meta-information). A web frontend could be used to provide an upload functionality for the company-owned library JARs, but the files needed for Maven are built using scripts and/or a Lean web application.

You pay a relatively low price for installing a Maven server, but you gain a huge amount of flexibility. Consider maintenance costs in this calculation alone. This chapter presents such a server, plus some scripts. It is up to you if you prefer to use standard products instead, but I think you should have the choice.

The Maven Repository Layout

If you look at one of the public Maven repositories, for example `https://repo.maven.apache.org/maven2/`, you will immediately get for example the following hierarchy:

```
..
org
  -> apache
    -> commons
      -> commons-math3
        -> 3.5
        -> 3.6
          -> [artifacts]
        -> ...
```

This directly corresponds to these Maven coordinates:

`org.apache.commons:commons-math3:3.6`

This means that, whenever Gradle decides to look into the Maven repository to resolve `org.apache.commons:commons-math3:3.6`, it will navigate to

```
https://repo.maven.apache.org/maven2/org/apache/commons/
        commons-math3/3.6/
```

To compile, for example, it will look for the following files in that URL:

```
.../commons-math3-3.6.jar
.../commons-math3-3.6.jar.asc
.../commons-math3-3.6.jar.md5
.../commons-math3-3.6.jar.sha1
.../commons-math3-3.6.pom
.../commons-math3-3.6.pom.asc
.../commons-math3-3.6.pom.md5
.../commons-math3-3.6.pom.sha1
```

where all the `.md5` and `.sha1` files are just the hashes that a client might or might not check to see whether the POM and JAR files contain the correct byte sequence. The `.pom` file defines transitive dependencies and other resolution-strategy and verification-related settings. The `.asc` file is optional and applies only to signed libraries. The `.jar` file contains the library itself.

It is important to know that the repository does not provide semantic functionality. It is up to the dependency resolution client (Gradle, Maven, Ivy, ...) to decide which files need to be downloaded. So, for a corporate Maven repository, omitting any signing, all you need is the `.jar` file and the adjacent `.jar.md5` and `.jar.sha1`.

Note   All those `*metadata.xml` files you see in the public repository are for list and search functionalities and do not affect resolution strategies.

For a corporate Maven repository, this chapter's strategy for providing libraries is as follows:

1. Check whether you can serve requests from a local (mirror) repository. If so, take the local version.

2. Otherwise, decide whether it is allowed to contact public repositories. This is optional. In the repository software, the answer will always be yes, but this is where you can hook in to comply with company policies.

3. If confirmed, fetch the libraries and POM files from the public repository.

4. Perform some checks (even manually by a human security appointee). This is optional.

5. If confirmed, save the library in the local (mirror) repository and deliver the library and POM to the requester.

In this example, I name it `Marvin`. I skip all the optional steps, but you are free to add them as extension work.

A Simple Server to Provide Mavenized Artifacts

You need to create a REST webserver that serves the Maven repository client. This consists of a root EAR Gradle project, one WAR Gradle child project, and one EJB Gradle child project.

The Marvin EAR Root Project

To start, create a Gradle project in Eclipse and name it Marvin. In the settings.gradle file, add the following:

```
include 'MarvinWeb', 'MarvinEjb'
```

This is to inform Gradle that you have two subprojects—MarvinWeb and MarvinEjb. Replace the contents of the build.gradle file with the following:

```
/*
 * GRADLE project build file
 */
plugins {
  id 'ear'
}

// !!!!!!!!!!!!!!!!!!!!!!!!!!!!!!!!!!!!!!!!!!!!!!!!!!!!!!!
// Use the deloyEar task to compile and assemble all and
// upload it to Glassfish
// !!!!!!!!!!!!!!!!!!!!!!!!!!!!!!!!!!!!!!!!!!!!!!!!!!!!!!!

repositories {
    jcenter()
}

dependencies {
    // The following dependencies will be the ear modules
    // and will be placed in the ear root
    deploy project(path: ':MarvinWeb',
        configuration: 'archives')
    deploy project(path: ':MarvinEjb',
        configuration: 'archives')

    // The following dependencies will become ear libs and
    // will be placed in a dir configured via the
    // libDirName property (default: 'lib')
    earlib 'org.apache.commons:commons-configuration2:2.6'
    earlib 'commons-io:commons-io:2.6'
```

```
    earlib 'commons-beanutils:commons-beanutils:1.9.4'
}

ear {
  // Possibly tweak the "ear" task. See the "ear" plugin
  // documentation
}

// This is a custom task to deploy the EAR on a local
// Glassfish instance
task deployEar(dependsOn: ear,
            description:">>> MARVIN deploy task") {
  doLast {
    def FS = File.separator
    def glassfish =
        project.properties['glassfish.inst.dir']
    def user = project.properties['glassfish.user']
    def passwd = project.properties['glassfish.passwd']

    File temp = File.createTempFile("asadmin-passwd",
        ".tmp")
    temp << "AS_ADMIN_${user}=${passwd}\n"

    def sout = new StringBuilder()
    def serr = new StringBuilder()
    def libsDir = "${project.projectDir}${FS}build" +
        "${FS}libs"
    def proc = """"${glassfish}${FS}bin${FS}asadmin
        --user ${user} --passwordfile ${temp.absolutePath}
        deploy --force=true ${libsDir}/${project.name}.ear
        """.execute()
    proc.waitForProcessOutput(sout, serr)
    println "out> ${sout}"
    if(serr.toString()) System.err.println(serr)

    temp.delete()

    new File(glassfish + FS +
```

```
      "glassfish${FS}domains${FS}domain1${FS}lib" +
      "${FS}classes" + FS + "marvin.properties").bytes =
    new File(project.rootDir.absolutePath + FS +
      "marvin.properties").bytes
  }
}

// This is a custom task for undeploying the application
// from a local Glassfish server
task undeployEar(
  description:">>> MARVIN undeploy task") {
  doLast {
    def FS = File.separator
    def glassfish =
        project.properties['glassfish.inst.dir']
    def user = project.properties['glassfish.user']
    def passwd = project.properties['glassfish.passwd']

    File temp = File.createTempFile("asadmin-passwd",
        ".tmp")
    temp << "AS_ADMIN_${user}=${passwd}\n"

    def sout = new StringBuilder()
    def serr = new StringBuilder()
    def proc = """${glassfish}${FS}bin${FS}asadmin
        --user ${user} --passwordfile ${temp.absolutePath}
        undeploy ${project.name}""".execute()
    proc.waitForProcessOutput(sout, serr)
    println "out> ${sout}"
    if(serr.toString()) System.err.println(serr)

    temp.delete()
  }
}
```

The following characteristics describe this file:

- The plugins { id 'ear' } loads and enables Gradle's EAR plugin.
  This plugin describes the construction of an .ear file.

- The `repositories { }` describes which repositories to contact for the necessary libraries.

- The `dependencies { }` section defines what needs to go into the EAR file. The `deploy` items tell about WARs an EJB should include; and the `earlib` items list library JARs to be included inside the EAR's `lib` folder. These dependency types are defined by the EAR plugin. You won't find them elsewhere.

- The two custom tasks are to create the Groovy scripts and to describe how to install and uninstall the EAR on a GlassFish server. They are not included in a build phase and therefore must be called explicitly. Obviously, these scripts must be rewritten for other server products.

- At the end of the deploy script, the application configuration file is copied to a place where the application can read it using classpath resolution. This procedure also depends on the server product in use.

- The operating system commands are executed inside both custom scripts. For Windows, just prepend `cmd /c` to the command strings.

In the project root, add a file called `gradle.properties`. Here, you see how to connect to the local GlassFish instance:

```
glassfish.inst.dir = /path/to/glassfish7
glassfish.user = admin
glassfish.passwd =
```

Add the `marvin.properties` file to the project root. This file contains the runtime configuration for Marvin:

```
repoDir = /path/to/mirror
externalRepos = http://repo.maven.apache.org/maven2
```

The `repoDir` specifies the path to your mirror repository. The `externalRepos` is a comma-separated list of external (/public) repositories to contact if the local mirror repository is missing any files.

The Marvin Web Project

For the WAR subproject, create a new top-level folder named MarvinWeb inside the EAR project and copy the gradlew, gradlew.bat, and gradle files and folders from the root project into it. This way gradle can be executed from inside the web-subproject using a terminal as well. I won't show that here, but you might want to use it, so it does not hurt having it here.

Inside the MarvinWeb folder, create a build.gradle file and add these contents:

```
/*
 * GRADLE project build file
 */
plugins {
  id 'war'
}
sourceCompatibility = 1.17
targetCompatibility = 1.17

ext {
  referencedEjbProjects = ['MarvinEjb']
}

repositories {
  jcenter()
}

dependencies {
  testImplementation 'junit:junit:4.13.2'
  implementation 'jakarta.platform:' +
      'jakarta.jakartaee-api:10.0.0'

  implementation 'org.apache.commons:' +
      'commons-configuration2:2.6'

  // No EJB project dependency here, because then we would
  // get the EJB implementations. Instead we only need the
  // interfaces
  referencedEjbProjects.each { ejb ->
    implementation files("${rootDir}/${ejb}/build/libs/" +
```

```
                            "ejb-interfaces.jar")
    }
}
```

This build file makes sure you use the WAR plugin, and it says to compile using Java 17. The `referencedEjbProjects` lists the EJB projects you depend on, which in this case is a single one named `MarvinEjb`. If you have more, you write `['MarvinEjb', 'OtherEjb', ...]`. For the dependencies, add the Jakarta EE API as a `compileOnly` dependency, meaning it won't be added to the WAR file. You also add the EJB interfaces, which the EJB projects create. I talk about that later.

Before you add the code for the web application, first declare the EJB project. This way, you can let Eclipse re-layout its project structure first, before you write the code for each subproject.

The Marvin EJB Project

For the EJB subproject, create a new top-level folder named `MarvinEjb` and copy the `gradlew`, `gradlew.bat`, and `gradle` files and folders from the root project into it. This again is allows Gradle for the EJB project to be run in a terminal.

Inside `MarvinEjb`, create a new build file called `build.gradle` and insert the following text:

```
/*
 * GRADLE project build file
 */
plugins {
  id 'java'
}
sourceCompatibility = 1.17
targetCompatibility = 1.17

repositories {
  jcenter()
}

dependencies {
  testImplementation 'junit:junit:4.13.2'
```

```
  implementation 'jakarta.platform:' +
      'jakarta.jakartaee-api:10.0.0'

  // Those need to be added in the root projects 'earlib'
  // dependencies:
  compileOnly 'org.apache.commons:commons-' +
        'configuration2:2.6'
  compileOnly 'commons-io:commons-io:2.6'
  compileOnly 'commons-beanutils:commons-' +
        'beanutils:1.9.4'
}

// Build a JAR with just the EJB interfaces,
// for clients
task('EjbInterfaces', type: Jar, dependsOn: 'classes') {
  //describe jar contents here as a CopySpec
  archiveFileName = "ejb-interfaces.jar"
  destinationDirectory = file("$buildDir/libs")
  from("$buildDir/classes/java/main") {
    include "**/ejb/interfaces/**/*.*"
  }
}
// make sure it is part of the assembly path
jar.dependsOn EjbInterfaces
```

The crucial part of this build script is that you added a custom task that builds an additional JAR containing the EJB interfaces. This is important, because clients, for example the web application, are supposed to use and bundle the EJB interface, *not* the implementations. This additional JAR is referred to in this special dependencies { } construct of the web application's build.gradle file.

Laying Out the Projects Again

Choose Gradle ➤ Refresh Gradle Project in the context menu of the root (EAR) project. Eclipse will show an extracted view with the two subprojects shown in the top-level hierarchy, as shown in Figure 6-1.

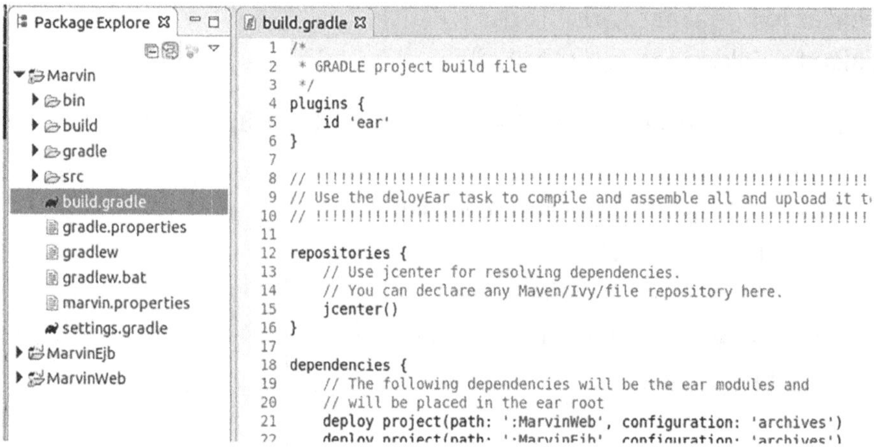

Figure 6-1. *Project laid out again*

Note   At this stage, the web subproject shows some errors. This is expected, since the EJB interfaces are not available yet. You learn to build them in the following sections.

The Web Project Code

Create a folder called `src/main/webapp/WEB-INF` inside the subproject and add the beans.xml file:

```
<?xml version="1.0" encoding="UTF-8"?>
<beans xmlns="https://jakarta.ee/xml/ns/jakartaee"
    xmlns:xsi="http://www.w3.org/2001/XMLSchema-instance"
    xsi:schemaLocation="https://jakarta.ee/xml/ns/jakartaee
        https://jakarta.ee/xml/ns/jakartaee/beans_3_0.xsd"
    bean-discovery-mode="all" version="3.0">
</beans>
```

Plus one file web.xml with contents

```
<?xml version="1.0" encoding="UTF-8"?>
<web-app xmlns:xsi=
```

```
     "http://www.w3.org/2001/XMLSchema-instance"
  xmlns="http://xmlns.jcp.org/xml/ns/javaee"
  xsi:schemaLocation="http://xmlns.jcp.org/xml/ns/javaee
     http://xmlns.jcp.org/xml/ns/javaee/web-app_4_0.xsd"
  id="WebApp_ID" version="4.0">
  <display-name>Marvin</display-name>
  <servlet>
    <servlet-name>
      jakarta.ws.rs.core.Application
    </servlet-name>
  </servlet>
  <servlet-mapping>
    <servlet-name>
      jakarta.ws.rs.core.Application
    </servlet-name>
    <url-pattern>/*</url-pattern>
  </servlet-mapping>
</web-app>
```

and another file glassfish-web.xml which reads

```
<?xml version="1.0" encoding="UTF-8"?>
<glassfish-web-app error-url="">
    <class-loader delegate="true"/>
</glassfish-web-app>
```

Create another folder called src/main/java/book/jakartapro/marvin/war and add src/main/java as the source folder by right-mouse-clicking (project) ➤ Properties ➤ Java Build Path ➤ Source. The implementation class reads Marvin:

```
package book.jakartapro.marvin.war;

import java.util.HashMap;
import java.util.Map;

import jakarta.ejb.EJB;
import jakarta.servlet.http.HttpServletRequest;
import jakarta.ws.rs.GET;
```

```java
import jakarta.ws.rs.Path;
import jakarta.ws.rs.PathParam;
import jakarta.ws.rs.core.Context;
import jakarta.ws.rs.core.Response;
import jakarta.ws.rs.core.Response.ResponseBuilder;

import book.jakartapro.marvin.ejb.interfaces.
        LocalMirrorLocal;

/**
 * REST Web Service
 */
@Path("/")
public class Marvin {
  static Map<String,String> CM = new HashMap<>();
  static {
    CM.put(".jar", "application/java-archive");
    CM.put(".pom", "text/xml");
  }

  @EJB LocalMirrorLocal localMirror;

  @GET
  @Path("repo/{p : .*}")
  public Response fetch(@PathParam("p") String path,
        @Context HttpServletRequest requ) {
    return getFromRepo(Response.status(200).
          entity(path), path).build();
  }

  private ResponseBuilder getFromRepo(ResponseBuilder rb,
        String path) {
    String suffix = path.substring(path.lastIndexOf("."));
    String outType = CM.getOrDefault(suffix, "text/plain");
    rb.type(outType);
    rb.entity(localMirror.fetchBytes(path, rb));
    return rb;
  }
}
```

The EJB Project Code

Inside the MarvinEjb sub-project, add a folder called src/main/java/book/jakartapro/ marvin/ejb and then add src/main/java as a source folder by right-mouse-clicking (project) ➤ Properties ➤ Java Build Path ➤ Source. Add a configuration helper class to the book/jakartapro/marvin/ejb package, as follows:

```java
package book.jakartapro.marvin.ejb;

import java.io.File;
import java.io.FileOutputStream;
import java.io.IOException;
import java.io.InputStream;

import org.apache.commons.configuration2.Configuration;
import org.apache.commons.configuration2.builder.fluent.
      Configurations;
import org.apache.commons.configuration2.ex.
      ConfigurationException;
import org.apache.commons.io.IOUtils;

public class Conf {
  private static Conf INSTANCE = null;
  public static Conf getInstance() {
  if(INSTANCE == null) INSTANCE = new Conf();
    return INSTANCE;
  }

  private Configuration c;

  private Conf() {
    try {
      Configurations configs = new Configurations();

      // From Class, the path is relative to the package
      // of the class unless you include a leading slash,
      // so if you don't want to use the current package,
      // include a slash like this:
      InputStream ins = this.getClass().
```

```
        getResourceAsStream("/marvin.properties");

    File f = File.createTempFile("marvin-x48", null);
    IOUtils.copy(ins, new FileOutputStream(f));
    c = configs.properties(f);
    f.delete();
  } catch (ConfigurationException | IOException e) {
    e.printStackTrace(System.err);
  }
}

public String string(String key) {
  return c.getString(key);
}
}
```

The sole purpose of this class is to simplify configuration access. The EJB implementation class is called LocalMirror. As its contents, write the following:

```
package book.jakartapro.marvin.ejb;

import java.io.File;
import java.io.FileInputStream;
import java.io.InputStream;
import java.net.HttpURLConnection;
import java.net.URL;

import jakarta.ejb.Local;
import jakarta.ejb.Stateless;
import jakarta.ws.rs.core.Response.ResponseBuilder;

import org.apache.commons.io.FileUtils;
import org.apache.commons.io.IOUtils;

import book.jakartapro.marvin.ejb.interfaces.
    LocalMirrorLocal;
```

```
@Stateless
@Local(LocalMirrorLocal.class)
public class LocalMirror {
  static Conf conf = Conf.getInstance();

  public byte[] fetchBytes(String path,
        ResponseBuilder rb) {
    String repoDir = conf.string("repoDir");
    File f = new File(repoDir + "/" + path);

    // If the file does not exist in our local (mirror)
    // repository, we contact the external (public)
    // Maven repository. This is the place where you
    // could add filters
    if (!f.exists())
      loadExternal(path, f, rb);

    // If it still does not exist, the public repo doesn't
    // have it either (and it was not copied into our
    // mirror repo), and we output an error message.
    // Otherwise read the bytes
    if (!f.exists()) {
      rb.type("text/plain");
      return ("Cannot load '" + path + "'\n").getBytes();
    } else {
      try {
        return IOUtils.toByteArray(
              new FileInputStream(f));
      } catch (Exception e) {
      }
      return null;
    }
  }

  /////////////////////////////////////////////////////////
  /////////////////////////////////////////////////////////
```

```java
  private void loadExternal(String path, File tgtFile,
        ResponseBuilder rb) {
    try {

      // Contact the external (public) repositories until
      // we find the file requested
      InputStream ins = null;
      String[] urls = conf.string("externalRepos").
            split(",");
      for (String r : urls) {
        HttpURLConnection c = (HttpURLConnection)
              new URL(r + "/" + path).openConnection();
        c.setRequestProperty("User-Agent", "Marvin");

        int respCode = c.getResponseCode();
        rb.status(respCode);

        if (respCode != 200)
          continue; // not yet found

        ins = c.getInputStream();
        break;
      }

      if (ins != null)
        writeToMirror(ins, tgtFile);
    } catch (Exception e) {
      e.printStackTrace(System.err);
    }
  }

  // Mirror the file in our local repo
  private void writeToMirror(InputStream ins,
        File tgtFile) throws Exception {
    tgtFile.getParentFile().mkdirs();
    FileUtils.copyInputStreamToFile(ins, tgtFile);
  }
}
```

Add the EJB interface `LocalMirrorLocal.class` inside the book/jakartapro/
marvin/ejbinterfaces package:

```
package book.jakartapro.marvin.ejb.interfaces;

import jakarta.ws.rs.core.Response.ResponseBuilder;

public interface LocalMirrorLocal {
  byte[] fetchBytes(String path, ResponseBuilder rb);
}
```

This interface will be needed by clients to use the EJB.

Building and Deploying the EAR

To build the EAR, double-click Marvin ➤ Build ➤ EAR inside the Gradle Tasks view. The
generated EAR file will be placed in the build/libs project.

Caution   Eclipse filters out the build folder from the Package Explorer view.
To see the folder, open the menu (find the menu button inside the taskbar) and
navigate to Filters. Then uncheck the Gradle Build Folder entry. Press F5 on that
folder to let Eclipse reread its contents.

To deploy the EAR file on the local server, select Marvin ➤ Other ➤ deployEar.
This task depends on the EAR task, so to build *and* deploy the deploy task is enough.
Obviously, the server must be running for this to work.

Caution   You must check Show All Tasks in the Gradle Task menu to see this
custom task.

Using the Corporate Repository

To use that corporate Maven repository in your build.gradle build files, enter the
following:

```
repositories {
  maven {
    url 'http://serv.er.addr:8080/MarvinWeb/repo
  }
}
```

Substitute the repository's server address. The port is the GlassFish HTTP connector's standard port. If you changed this, you have to change the port number accordingly.

Note   To perform a quick check in your browser, enter this for example:

```
http://serv.er.addr:8080/MarvinWeb/repo/junit/junit/4.13.2/
junit-4.13.2.pom
```

A Sample Java Library

As an example, you can write a simple library Gradle project with one Java class that counts word frequencies in some arbitrary text. Start a new Gradle project and replace the contents of the build.gradle file with the following:

```
plugins {
    id 'java-library'
}
sourceCompatibility = 1.17
targetCompatibility = 1.17

repositories {
    jcenter()
}

dependencies {
    api 'org.apache.commons:commons-math3:3.6.1'
    implementation 'com.google.guava:guava:28.0-jre'
    testImplementation 'junit:junit:4.13.2'
}
```

The class reads as follows:

```
package book.jakartapro.textutils;

import java.util.Comparator;
import java.util.List;
import java.util.stream.Collectors;

import com.google.common.base.CharMatcher;
import com.google.common.base.Splitter;
import com.google.common.collect.HashMultiset;
import com.google.common.collect.Multiset;

public class TextUtils {
  public static class Pair<S,T> {
    public S s;
    public T t;
    public Pair(S s, T t) {
      this.s = s;
      this.t = t;
    }
    public S first() { return s;}
    public T second() { return t; }
    @Override public String toString() {
        return "(" + s + "," + t + ")"; };
  }

  public static List<Pair<String,Integer>>
      wordFrequenciesSorted(String text) {
    Multiset<String> mus = Splitter.
      onPattern("[^\\p{L}\\p{Nd}]+").
      trimResults(CharMatcher.anyOf(
                ".,;:-'\"!?=()[]{}*/")).
      omitEmptyStrings().
      splitToList(text).stream().collect(
          () -> { return HashMultiset.create(); },
        (ms,str) -> { ms.add(str.toLowerCase()); },
```

```
           (ms1,ms2) -> { ms1.addAll(ms2); });
    return mus.entrySet().stream().
      sorted(Comparator.comparing((e) -> {
        return String.format("%09d-%s",
              10000000-e.getCount(),
              e.getElement());
      })).
      map((Multiset.Entry<String>e) -> {
        return new Pair<String,Integer>(e.getElement(),
                                e.getCount()); }).
      collect(Collectors.toList());
  }
}
```

You can see that this example uses the `java-library` plugin, not `java`. The difference is that the former introduces two dependency types—`implementation` and `api`—and the latter describes a dependency transitively forwarded to client projects referring to the library. In contrast, `implementation` is needed only internally by the library.

You can then push all files to a Git repository, and afterward you can invoke the `toRepo.sh` script that I talk about in the following section to copy the library to the Maven repository.

Building and Uploading Company Libraries

Since you know that, to upload a library to the company's Maven repository, you need the following:

- The library JAR itself

- A `.pom` file describing the library and its dependencies

- MD5 and SHA1 files of all of the above

You have to find a way to generate a POM file and the hashes. A script to automate this process as much as possible follows:

1. Check out the project from a repository.

2. Build the project (this means create the library JAR).

3. Generate a POM file by an adapted `build.gradle` file.

4. Add MD5 and SHA1 hashes from the same build file.

5. Upload everything to the company's Maven repository.

The first thing you do is create a script file which, by virtue of an included Groovy script, creates an adapted Gradle build file. This script file could be a bash script, but this would be rather complex because you need some replacement activities, which in bash is not easy to do. Instead, you use a Groovy script. Since you have Gradle at hand and Gradle has a complete Groovy inside, you can pack the script into a Gradle build file. To this aim, create an empty directory and add a `buildRepoGradle.gradle` file with the following contents:

```
task a { doLast {
  def FS = File.separator
  def buildFile = new File("build.gradle")
  def newBuild = buildFile.readLines().
    grep { ln -> !( ln ==~ /\s*archivesBaseName\s*=.*/ ) }.
    grep { ln -> !( ln ==~ /\s*group\s*=.*/ ) }.
    grep { ln -> !( ln ==~ /\s*version\s*=.*/ ) }.
    grep { ln -> !( ln ==~ /\s*id\s+('|")maven("|')\s*/) }.
    join("\n")

  // Add the maven plugin. This is able to generate POM
  // files
  newBuild = newBuild.replaceFirst(
      /(?s)plugins\s*\{([^}]*)\}/) { all, p ->
    "plugins {" + p + "  id 'maven'\n" + "}"
  }

  // Add Maven coordinates
  newBuild = newBuild.replaceFirst(
      /(?s)plugins\s*\{[^}]*\}/) { all ->
    all +
        "\ngroup = '" +
            project.properties.GROUP_ID + "'" +
        "\narchivesBaseName = '" +
            project.properties.ARTIFACT_ID + "'" +
```

```
            "\nversion = '" +
                project.properties.VERSION + "'" +
            "\n"
  }

  // Add the POM generation task
  newBuild += """\n
  task writePomAndHashes() {
    doLast {
      def FS = File.separator
      def jarName = '""" + project.properties.ARTIFACT_ID +
        "-" + project.properties.VERSION + ".jar" + """'
      def pomName = '""" + project.properties.ARTIFACT_ID +
        "-" + project.properties.VERSION + ".pom" + """'

      pom {
        project {
          inceptionYear '2019'
          licenses {
            license {
              name 'Property of the company'
            }
          }
    }
      }.writeTo(buildDir.absolutePath + FS +
                "poms" + FS + pomName)

      // Add the hashes
      def jarFile = buildDir.absolutePath + FS +
        'libs' + FS + jarName
      def pomFile = buildDir.absolutePath + FS +
        'poms' + FS + pomName
      def md5 = java.security.MessageDigest.
        getInstance("MD5")
      def sha1 = java.security.MessageDigest.
        getInstance("SHA-1")
      new File(jarFile+".md5").bytes =
```

```
        md5.digest(jarFile.bytes)
      new File(jarFile+".sha1").bytes =
        sha1.digest(jarFile.bytes)
      new File(pomFile+".MS5").bytes =
        md5.digest(pomFile.bytes)
      new File(pomFile+".sha1").bytes =
        sha1.digest(pomFile.bytes)

      // Write an info file
      def infoFile = new File(buildDir.absolutePath + FS +
        "repoFiles.txt")
      infoFile.delete()
      infoFile << jarFile + "\\n" +
          jarFile + ".md5\\n" +
          jarFile + ".sha1\\n" +
          pomFile + "\\n" +
          pomFile + ".md5\\n" +
          pomFile +".sha1\\n"
    }
  }"""

  new File("build2.gradle").text = newBuild
} }
```

The script takes the original build.gradle file and first removes some declarations
that you'll do yourself, like setting the Maven artifact ID, the Maven group ID, and
the Maven version. Then make sure the Maven plugin is loaded, since this plugin can
generate the POM file you need. Then set the Maven project coordinates and write the
POM generator task, which includes calculating the hashes. The last thing the task does
is create an info file, which lists the generated artifacts.

Next is a bash script toRepo.sh. It checks out the library project, invokes this Gradle
generator script, and then copies the generated artifacts to the Maven repository:

```
#!/bin/bash

GROUP_ID=book.jakartapro
ARTIFACT_ID=textutils
```

```
VERSION=1.0
REPO_BASE=`pwd`/REPO
GIT_URL=/the/path/to/GITREPO/TextUtils

# Checkout from GIT
rm -rf checkout && mkdir checkout
cd checkout
git clone $GIT_URL .
cp ../buildRepoGradle.gradle .

# -------------------------------------------------
# Create a modified build.gradle file: build2.gradle
./gradlew -b buildRepoGradle.gradle -q \
    -PGROUP_ID=${GROUP_ID} \
    -PARTIFACT_ID=${ARTIFACT_ID} \
    -PVERSION=${VERSION} \
a
# -------------------------------------------------

./gradlew -b build2.gradle build
./gradlew -b build2.gradle writePomAndHashes
cd ..

# Copy everything into the repository
pathInRepo=\
    ${GROUP_ID//\./\/}/${ARTIFACT_ID//\./\/}/${VERSION}
mkdir -p $REPO_BASE/$pathInRepo
while read p; do
  cp $p $REPO_BASE/$pathInRepo
done < checkout/build/repoFiles.txt

rm -rf checkout
```

Adapt the variables on top of the file and then invoke the script: `./toRepo.sh`. Make sure first the file is executable (`chmod a+x toRepo.sh`). The Git URL can of course also be a network URL, like `http://...` or `ssh://...`, but if you are on the same machine, you can also just enter the path. Git knows that you mean the `file://...` protocol.

PART II

Advanced Web Tier Topics

CHAPTER 7

Facelets

This chapter covers using Faces 4.0's built-in templating engine, called *Facelets*. Compared to using plain Faces, you can use Facelets to further modularize your frontend code and improve code quality, reuse, and maintenance.

Note The old name of Faces, *JSF*, is no longer used.

Faces Templating via Facelets

While you can create a full page-flow web application by using various fully defined Faces (XHTML) pages and the tags from the following component namespaces, it is probably not the cleverest way to rely on those three tag libraries:

```
h="jakarta.faces.html"
f="jakarta.faces.core"
pt="jakarta.faces.passthrough"
```

(The last is a pseudo-component namespace for attributes *not* to be handled by Faces.) If you have a common page structure with, for example, a header, a menu, a footer, and a content area, all Faces pages have to repeat all the page elements that are common to the different application parts. This is a violation of the DRY (Don't Repeat Yourself) principle. Instead, it would be better if you could use templating mechanisms to compose pages, given patterns, (*X*)HTML snippets, and placeholders.

Faces includes such a templating technology, and it is called *Facelets*. Facelets allows you to introduce parameterized template HTML pages, HTML snippets (components) to be included in pages, placeholders for such snippets, and decorators and repetitions

© Peter Späth 2023
P. Späth, *Pro Jakarta EE 10*, https://doi.org/10.1007/978-1-4842-8214-4_7

for things like elaborated list views. In the following sections, you learn about the configuration issues and about the Facelets tags. In the last section of this chapter, you develop a sample application to get you started.

This sample application mimics the functionality of an online music box. It includes a header, a footer, and a menu that appears on every page of the web application, no matter which functionality the user is currently using. Facelets does a good job at letting you factor out common page parts so you have to code them only once. See Figure 7-1.

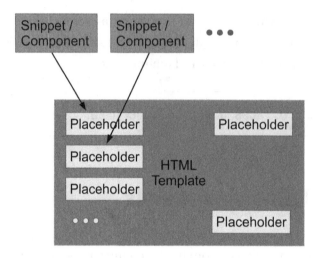

Figure 7-1. *Templating with Facelets*

Installing Facelets

Facelets is already part of Faces and Jakarta EE, so there is no need to install it. But before you can use it in your Faces pages, you have to declare an additional namespace, as the ui in the following:

```
<!DOCTYPE html>
<html xmlns:h="jakarta.faces.html"
  xmlns:f="jakarta.faces.core"
  xmlns:ui="jakarta.faces.facelets"
  xmlns:pt="jakarta.faces.passthrough">
    ...
</html>
```

Facelets Tags Overview

Facelets consists of the following tags, which you can include in an XHTML file to apply or mix templates, include XHTML snippets, or pass parameters:

The <ui:include> Tag

Include another XHTML file, as follows:

```
<ui:include src="incl.xhtml" />
```

If the file contains a <ui: composition> or a <ui:component>, only the inner contents of the <ui:composition> or <ui:component> will be included. This allows designers to style the included files independent of their later plumbing together by the server.

First Variation of the <ui:composition> Tag

If this is used *without* template="...", as in the following:

```
<ui:composition>
   ...
</ui:composition>
```

It defines a subtree (collection) of HTML elements. The idea behind that is, if you use <ui:include> and the file included contains a <ui:composition> ... </ui:composition>, only the inner contents of the <ui:composition> ... </ui:composition> will be included. The tag itself and anything around it will be ignored. So you can let page designers create a completely valid XHTML file, put a <ui:composition> ... </ui:composition> around interesting parts, and write <ui:include> in any other Faces pages to extract exactly such parts.

Second Variation of the <ui:composition> Tag

If it's used *with* template="...", as in the following:

```
<ui:composition template="templ.xhtml">
    ...
</ui:composition>
```

It defines a collection of XHTML snippets to be passed into placeholders inside the template file (corresponding to the `template = "..."` attribute).

This is a completely different usage scenario compared to `<ui:composition>` without `template="..."`. In the template file, you have one or more elements, such as `<ui:insert name="name1" />`, and in the file with the `<ui:composition template="...">`, you use `<ui:define>` tags *inside* the `<ui:composition template="..."> ... </ui:composition>`, as follows:

```
<ui:composition template="templ.xhtml">
    <ui:define name = "someName"> ... </ui:define>
    <ui:define name = "someName2"> ... </ui:define>

    ...
</ui:composition>
```

This defines contents to be used for the `<ui:insert>` tags. Anything around the `<ui:composition>` tag will be ignored again, so you can let designers create the snippets using non-Faces-aware HTML editors and only later extract interesting parts with `<ui:define name = "someName"> ... </ui:define>`, to be used for materializing the template file.

The <ui:insert> Tag

Use this tag to define placeholders inside template files. Using this tag inside a template file means that any file referring to this template may define contents for the placeholders:

```
<ui:insert name="name1"/>
```

This definition has to happen inside of `<ui:composition>`, `<ui:component>`, `<ui:decorate>`, or `<ui:fragment>`.

Usually you don't provide contents in this tag. If however you add contents, such as:

```
<ui:insert name="name1">
    Hello
</ui:insert>
```

It will be taken as a default if the placeholder is not otherwise defined.

The <ui:define> Tag

This tag declares what will be inserted at the insertion points:

```
<ui:define name="theName">
    Contents...
</ui:define>
```

Since insertion points can only exist in template files, the <ui:define> tag can only show up in files referring to template files via <ui:composition template = "...">.

The <ui:param> Tag

Specify a parameter that's passed to an <ci:include>-ed file, or to the template specified in <ui:composition template = "..."> ... Simply add as a child element, such as:

```
<ui:include src="comp1.xhtml">
    <ui:param name="p1" value="Mark" />
</ui:include>
```

Inside the referred-to file, use #{paramName} to use the parameter:

```
<h:outputText value="Hello #{p1}" />
```

The <ui:component> Tag

This is the same as the first variation of <ui:composition>, without template specification. It adds an element to the Faces component tree. It supports the following attributes:

- id for the element's ID in the component tree. It's not required, and Faces generates an automatic ID if you don't specify it. May be an EL (expression language) string value.

- binding for binding the component to a Java class (must inherit from jakarta.faces.component.UIComponent). Not required. May be an EL string value (class name).

- rendered for determining whether the component is to be rendered. Not required. May be an EL boolean value.

It is common practice to use `<ui:param>` to pass parameters to components. You can for example tell the component to use a particular ID. Caller:

```
<ui:include src="comp1.xhtml">
    <ui:param name="id" value="c1" />
</ui:include>
```

Callee (`comp1.xhtml`):

```
<ui:component id="#{id}">
  ...
</ui:component>
```

The <ui:decorate> Tag

Similar to `<ui:composition>`, but it does *not* disregard the XHTML code around it:

```
...
I'm written to the output!
<ui:decorate template="templ.xhtml">
    <ui:define name="def1">
        I'm passed to "templ.xhtml", you can refer to
        me in "templ.xhtml" via
        &lt;ui:insert name="def1"/&gth;
    </ui:define>
</ui:include>
...
```

In contrast to `<ui:composition>`, the file with the `<ui:decorate>` will contain the completely valid XHTML code, including `html`, `head`, and `body`, and the template file will be inserted where the `<ui:decorate>` appears. It therefore must not contain `html`, `head`, or `body`. This is thus more or less an extended "include," where passed-over data is not given by the attributes but in the `body` tag instead.

You usually apply the `<ui:decorate>` tag to further elaborate code snippets. You can wrap them into more `<div>`s to apply more styles, add a label or a heading, and more.

The <ui:fragment> Tag

Similar to <ui:decorate>, but this tag creates an element in the Faces component tree. It has the following attributes:

- id for the element's ID in the component tree. This is not required, and Faces generates an automatic ID if you don't specify it. May be an EL (expression language) string value.

- binding for binding the component to a Java class (must inherit from jakarta.faces.component.UIComponent). Not required. May be an EL string value (class name).

- rendered for indicating whether the component is to be rendered. Not required. May be an EL boolean value.

You can use this tag to extract existing code snippets and to convert them partly to a component. For example, consider the following code:

```
<DOCTYPE html>
<html ...><head>...</head>
<h:body>
  ...
  <table>
      [Some table|
  </table>
  ...
</h:body></html>
```

If you now extract the table to a different file, called table1_frag.xhtml:

```
<!-- Caller: ############################ -->
<!-- original file                        -->
<DOCTYPE html>
<html ...><head>...</head>
<h:body>
  ...
  <ui:include src="table1_frag.xhtml"/>
  ...
</h:body></html>
```

95

```
<!-- Callee: ########################### -->
<!-- table1_frag.xhtml                -->
<div xmlns="http://www.w3.org/1999/xhtml"
    xmlns:h="jakarta.faces.html"
xmlns:f="jakarta.faces.core"
xmlns:ui="jakarta.faces.facelets"
xmlns:pt="jakarta.faces.passthrough">
<div>I am the table caption</div>
<ui:fragment>
  <table>
    [Some table|
  </table>
</ui:fragment>
</div>
```

You have introduced XHTML (the caption) and a new component (the table).

The <ui:repeat> Tag

This is not a necessarily templating-related tag, but it's used to loop over a collection or an array. Its attributes are:

- begin. Not required. If it's specified, it's the iteration's first index in the list or array. May be an int valued value expression.

- end. Not required. If it's specified, it's the iteration last index (inclusive) in the list or array. May be an int valued value expression.

- step. Not required. If it's specified, it's the stepping that inside the list or array. May be an int valued value expression.

- offset. Not required. If it's specified, it's an offset that's added to the iterated-over values. May be an int valued value expression.

- size. Not required. If it's specified, it is the maximum number of elements to read from the collection or array. Must not be greater than the collection's array size.

- value. The list or array to iterate over. An Object valued expression. Required.

- var. The name of an expression language variable that holds the current item of the iteration. May be a String value expression.

- varStatus. Not required. The name of a variable that holds the iteration status. A POJO with read-only values: begin (int), end (int), index (int), step (int), even (boolean), odd (boolean), first (boolean), and last (boolean).

- rendered. Whether or not the component is to be rendered. Not required. May be an EL boolean value.

Note   The JSTL (Java Standard Tag Library) collection provides a <c:forEach> tag for looping. Faces and JSTL do not work together very well because of conceptional differences. In tutorials and blogs, you will find lots of examples of for loops with JSTL. It is, however, better to use <ui:repeat> to avoid problems.

The <ui:debug> Tag

Add this tag to your page during the development phase of your project. Using a hotkey, the tag will then lead to the Faces component tree and other information to be shown on the page. Use the hotkey="x" attribute to change the hotkey. Shift+Ctrl+X will then display the component (the default *d* does not work with the Firefox browser!). The second optional attribute is rendered="true|false" (you can also use an EL boolean expression), which switches this component on and off.

Note   This tag works only in the development project stage. Inside web.xml, you can add the following

```
<context-param>
    <param-name>jakarta.faces.PROJECT_STAGE</param-name>
    <param-value>Development</param-value>
</context-param>
```

to specify the project stage (Development (default), UnitTest, SystemTest, or Production).

An Example Facelets Project

In this section, you learn how to build an example Facelets project with a music box database showing similarly designed pages for titles, composers, and performers. Start a new Gradle project in Eclipse and name it MusicBox. Take the build.gradle file and replace its contents with the following:

```
plugins {
  id 'war'
}
sourceCompatibility = 1.17
targetCompatibility = 1.17

repositories {
  jcenter()
}

dependencies {
  compileOnly 'jakarta.platform:jakarta.jakartaee-api:10.0.0'
  implementation 'com.google.guava:guava:31.1-jre'
  testImplementation 'junit:junit:4.13.2'
}

task deployWar(dependsOn: war,
               description:">>> MUSICBOX deploy task") {
  doLast {
    def FS = File.separator
    def pp = { a -> project.properties[a] }
    def glassfish = pp 'glassfish.inst.dir'
    def user = pp 'glassfish.user'
    def passwd = pp 'glassfish.passwd'

    File temp = File.createTempFile("asadmin-passwd",
        ".tmp")
    temp << "AS_ADMIN_${user}=${passwd}\n"

    def sout = new StringBuilder()
    def serr = new StringBuilder()
    def libsDir =
```

```
          "${project.projectDir}${FS}build${FS}libs"
      def proc = """"${glassfish}${FS}bin${FS}asadmin
        --user ${user} --passwordfile ${temp.absolutePath}
        deploy --force=true ${libsDir}/${project.name}.war
      """.execute()
      proc.waitForProcessOutput(sout, serr)
      println "out> ${sout}"
      if(serr.toString()) System.err.println(serr)
      temp.delete()
    }
}

task undeployWar(
        description:">>> MUSICBOX undeploy task") {
  doLast {
    def FS = File.separator
    def pp = { a -> project.properties[a] }
    def glassfish = pp 'glassfish.inst.dir'
    def user = pp 'glassfish.user'
    def passwd = pp 'glassfish.passwd'

    File temp = File.createTempFile("asadmin-passwd",
        ".tmp")
    temp << "AS_ADMIN_${user}=${passwd}\n"

    def sout = new StringBuilder()
    def serr = new StringBuilder()
    def proc = """"${glassfish}${FS}bin${FS}asadmin
      --user ${user} --passwordfile ${temp.absolutePath}
      undeploy ${project.name}""".execute()
    proc.waitForProcessOutput(sout, serr)
    println "out> ${sout}"
    if(serr.toString()) System.err.println(serr)

    temp.delete()
  }
}
```

Apart from dependency handling, this build file introduces two custom tasks for deploying and undeploying the MusicBox web application on a local server.

To connect to the asadmin tool, create another file, called gradle.properties, in the project root:

```
glassfish.inst.dir = /path/to/glassfish7
glassfish.user = admin
glassfish.passwd =
```

where you enter your own GlassFish server installation path. An empty admin password is the default setting for GlassFish. If you changed this default setting, you must enter the password in this file.

For the music box data, you'll create three Java classes. For simplicity, they return static information. In real life, you would connect to a database to get the data. Create a package called book.jakartapro.musicbox and add the following:

```java
// Composers.java:
package book.jakartapro.musicbox;

import java.io.Serializable;
import java.util.List;

import jakarta.enterprise.context.SessionScoped;
import jakarta.inject.Named;

import com.google.common.collect.Lists;

@SessionScoped
@Named
public class Composers implements Serializable {
    private static final long serialVersionUID =
        -5244686848723761341L;

    public List<String> getComposers() {
        return Lists.newArrayList("Brahms, Johannes",
            "Debussy, Claude");
    }
}
```

```java
// Titles.java:
package book.jakartapro.musicbox;

import java.io.Serializable;
import java.util.List;

import jakarta.enterprise.context.SessionScoped;
import jakarta.inject.Named;

import com.google.common.collect.Lists;

@SessionScoped
@Named
public class Titles implements Serializable {
    private static final long serialVersionUID =
        -1034755008236485058L;

    public List<String> getTitles() {
        return Lists.newArrayList("Symphony 1",
            "Symphony 2", "Childrens Corner");
    }
}
```

```java
// Performers.java:
package book.jakartapro.musicbox;

import java.io.Serializable;
import java.util.List;

import jakarta.enterprise.context.SessionScoped;
import jakarta.inject.Named;

import com.google.common.collect.Lists;

@SessionScoped
@Named
public class Performers implements Serializable {
    private static final long serialVersionUID =
        6941511768526140932L;
```

```
public List<String> getPerformers() {
    return Lists.newArrayList(
        "Gewandhausorchester Leipzig",
        "Boston Pops");
    }
}
```

Although you won't internationalize this sample application for the sake of brevity, you do prepare internationalization and create an empty WebMessages.properties file inside src/main/resources/book/jakartapro/musicbox/web.

For Faces to work correctly, you need to create a src/main/webapp/WEB-INF/beans. xml file:

```
<?xml version="1.0" encoding="UTF-8"?>
<beans xmlns="https://jakarta.ee/xml/ns/jakartaee"
    xmlns:xsi="http://www.w3.org/2001/XMLSchema-instance"
    xsi:schemaLocation="https://jakarta.ee/xml/ns/jakartaee
        https://jakarta.ee/xml/ns/jakartaee/beans_3_0.xsd"
    bean-discovery-mode="all" version="3.0">
</beans>
```

Another file, called src/main/webapp/WEB-INF/web.xml, must read as follows:

```
<?xml version="1.0" encoding="UTF-8"?>
<web-app xmlns:xsi=
  "http://www.w3.org/2001/XMLSchema-instance"
  xmlns="http://xmlns.jcp.org/xml/ns/javaee"
  xsi:schemaLocation="http://xmlns.jcp.org/xml/ns/javaee
    http://xmlns.jcp.org/xml/ns/javaee/web-app_4_0.xsd"
  id="WebApp_ID" version="4.0">

  <display-name>MusicBox</display-name>
  <welcome-file-list>
    <welcome-file>start.xhtml</welcome-file>
  </welcome-file-list>
```

```
<servlet>
  <servlet-name>Faces Servlet</servlet-name>
  <servlet-class>
    jakarta.faces.webapp.FacesServlet
  </servlet-class>
  <load-on-startup>1</load-on-startup>
</servlet>
<servlet-mapping>
  <servlet-name>Faces Servlet</servlet-name>
  <url-pattern>*.xhtml</url-pattern>
</servlet-mapping>
</web-app>
```

Add one more file, called src/main/webapp/WEB-INF/glassfish-web.xml. It should contain the following:

```
<?xml version="1.0" encoding="UTF-8"?>
<glassfish-web-app error-url="">
    <class-loader delegate="true"/>
</glassfish-web-app>
```

This last configuration file, called src/main/webapp/WEB-INF/faces-config.xml, defines the language files:

```
<?xml version="1.0" encoding="UTF-8"?>
<faces-config
    xmlns="https://jakarta.ee/xml/ns/jakartaee"
    xmlns:xsi="http://www.w3.org/2001/XMLSchema-instance"
    xsi:schemaLocation="https://jakarta.ee/xml/ns/jakartaee
        https://jakarta.ee/xml/ns/jakartaee/
        web-facesconfig_4_0.xsd"
    version="4.0">

<application>
  <resource-bundle>
    <base-name>
      book.jakartapro.musicbox.web.WebMessages
    </base-name>
```

```
    <var>bundle</var>
  </resource-bundle>
  <locale-config>
    <default-locale>en</default-locale>
    <!-- <supported-locale>es</supported-locale> -->
  </locale-config>
</application>
</faces-config>
```

Before you get into the coding, see Figure 7-2 to get an impression of what you actually want to achieve. To apply the Facelets functionalities, you need to add one template file, called src/main/webapp/frame.xhtml:

```
<!DOCTYPE html>
<html xmlns:h="jakarta.faces.html"
    xmlns:f="jakarta.faces.core"
    xmlns:ui="jakarta.faces.facelets"
    xmlns:pt="jakarta.faces.passthrough">
<h:head>
  <title>Musicbox</title>
  <h:outputStylesheet library="css" name="style.css" />
</h:head>

<h:body>
    <div class="header-line">
        <ui:insert name="header"><h2>Top Section</h2></ui:insert>
    </div>
    <div class="center-line">
      <div class="menu-column">
        <ui:insert name="menu"><ul><li>Menu1</li><li>Menu2</li></ul>
        </ui:insert>
      </div>
      <div class="contents-column">
        <ui:insert name="contents">Contents</ui:insert>
        <h:messages />
      </div>
```

```
    </div>
    <div class="bottom-line">
        <ui:insert name="footer">Footer</ui:insert>
    </div>
</h:body>
</html>
```

This template file defines a common page structure and declares a couple of placeholders via <ui:insert> tags. The CSS file is called style.css and it goes to src/main/webapp/resources/css/style.css:

```
body { color: blue; }
.header-line { height: 3em; background-color: #CCF000; }
.bottom-line { clear: both; height: 1.5em; }
.menu-column { float: left; width: 8em;
    background-color: #FFC000; height: calc(100vh - 7em); }
.menu-column ul { margin:0.5em; padding: 0;
    list-style-position: inside; }
.contents-column { float: left; padding: 0.5em;
    background-color: #FFFF99;
    width: calc(100% - 9em); height: calc(100vh - 8em); }
.bottom-line { padding-top: 1em;
    background-color: #CCFFFF; }
```

For common page elements, you define a couple of XHTML files inside the src/main/webapp/common folder:

```
<!-- File commonHeader.xhtml -->
<!DOCTYPE html>
<div xmlns:h="jakarta.faces.html"
     xmlns:f="jakarta.faces.core"
     xmlns:ui="jakarta.faces.facelets"
     xmlns:pt="jakarta.faces.passthrough">
  <h2>Musicbox</h2>
</div>

<!-- File commonMenu.xhtml -->
```

```
<!DOCTYPE html>
<div xmlns:h="jakarta.faces.html"
     xmlns:f="jakarta.faces.core"
     xmlns:ui="jakarta.faces.facelets"
     xmlns:pt="jakarta.faces.passthrough">

  <ul>
    <li><a href="titles.xhtml">Titles</a></li>
    <li><a href="composers.xhtml">Composers</a></li>
    <li><a href="performers.xhtml">Performers</a></li>
  </ul>
</div>

<!-- File commonFooter.xhtml -->
<!DOCTYPE html>
<div xmlns:h="jakarta.faces.html"
     xmlns:f="jakarta.faces.core"
     xmlns:ui="jakarta.faces.facelets"
     xmlns:pt="jakarta.faces.passthrough">

  (c) The Musicbox company 2019
</div>
```

The three page files inside the src/main/webapp folder—titles.xhtml, composers.
xhtml, and performers.xhtml—refer to the template file and the common page
elements:

```
<!-- File titles.xhtml ********************** -->
<html xmlns:h="jakarta.faces.html"
      xmlns:f="jakarta.faces.core"
      xmlns:ui="jakarta.faces.facelets"
      xmlns:pt="jakarta.faces.passthrough">
<h:body>
<ui:composition template="frame.xhtml">

  <ui:define name="header">
    <ui:include src="common/commonHeader.xhtml" />
  </ui:define>
```

```
<ui:define name="menu">
  <ui:include src="common/commonMenu.xhtml" />
</ui:define>

<ui:define name="contents">
  <h2>Titles</h2>
  <ul>
    <ui:repeat var="t" value="#{titles.titles}"
          varStatus="status">
      <li><h:outputText value="#{t}" /></li>
    </ui:repeat>
  </ul>
</ui:define>

<ui:define name="footer">
  <ui:include src="common/commonFooter.xhtml" />
</ui:define>

</ui:composition>
</h:body>
</html>

<!-- File composers.xhtml ********************* -->
<html xmlns:h="jakarta.faces.html"
    xmlns:f="jakarta.faces.core"
    xmlns:ui="jakarta.faces.facelets"
    xmlns:pt="jakarta.faces.passthrough">
<h:body>
<ui:composition template="frame.xhtml">

  <ui:define name="header">
    <ui:include src="common/commonHeader.xhtml" />
  </ui:define>

  <ui:define name="menu">
    <ui:include src="common/commonMenu.xhtml" />
  </ui:define>
```

```
<ui:define name="contents">
  <h2>Composers</h2>
  <ul>
    <ui:repeat var="c" value="#{composers.composers}"
          varStatus="status">
      <li><h:outputText value="#{c}" /></li>
    </ui:repeat>
  </ul>
</ui:define>

<ui:define name="footer">
  <ui:include src="common/commonFooter.xhtml" />
</ui:define>

</ui:composition>
</h:body>
</html>

<!-- File performers.xhtml ********************* -->
<html xmlns:h="jakarta.faces.html"
    xmlns:f="jakarta.faces.core"
    xmlns:ui="jakarta.faces.facelets"
    xmlns:pt="jakarta.faces.passthrough">
<h:body>
<ui:composition template="frame.xhtml">

  <ui:define name="header">
    <ui:include src="common/commonHeader.xhtml" />
  </ui:define>

  <ui:define name="menu">
    <ui:include src="common/commonMenu.xhtml" />
  </ui:define>

  <ui:define name="contents">
    <h2>Performers</h2>
    <ul>
```

```
    <ui:repeat var="p" value="#{performers.performers}"
            varStatus="status">
        <li><h:outputText value="#{p}" /></li>
    </ui:repeat>
    </ul>
  </ui:define>

  <ui:define name="footer">
    <ui:include src="common/commonFooter.xhtml" />
  </ui:define>

</ui:composition>
</h:body>
</html>
```

Note that the `<ui:composition>` tag applies the page template.

Build and deploy the application by running the Gradle task called `deployWar`. Then point your browser to the following URL to see the running application:

```
http://localhost:8080/MusicBox
```

See Figure 7-2.

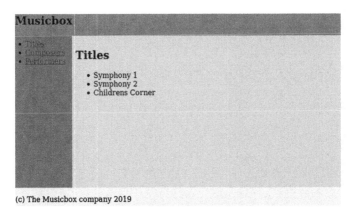

Figure 7-2. *The Musicbox Facelets application*

CHAPTER 8

Faces Custom Components

Custom components allow developers to design their own Faces components. The key point is usability—a custom component may easily be transported to other projects, or you might even decide to build a company-wide custom component collection.

There are basically three ways to define custom components. First, you can define tag files that you then register in `web.xml` as your own custom namespace. Second, you can use the *composite component* feature of Faces. Here, you use only XHTML files to define interfaces and implementations of components. Third, you can use custom Java classes to define components. The following sections describe all three of these methods.

Custom Tag Libraries

Here, you define your own XML namespace for use by XHTML pages, and this example uses components defined in a special XHTML file. The steps to create a custom tag library are as follows:

1. Create a component definition file called `mycomponent.xhtml` and place it in the `src/main/webapp/WEB-INF/tags` folder (create the folder if it doesn't exist). The name is free, but write it down since you will need it for the configuration files.

2. Create a tag library descriptor file called `mytaglib.taglib.xml`. The name is free again, but it goes to the configuration files as well. You describe the new component in this file.

3. Register the new tag library descriptor using `web.xml`.

© Peter Späth 2023
P. Späth, *Pro Jakarta EE 10*, https://doi.org/10.1007/978-1-4842-8214-4_8

As an example, you will create a simple date input component. The example doesn't add graphical goodies like a popup calendar, but you could later extend it to do such things. Here, you'll only take care of adding the correct user input in the form YYYY-MM-DD.

Create a file called fancyDateInput.xhtml inside src/main/webapp/WEB-INF/tags and let it read as follows:

```
<html xmlns:h="jakarta.faces.html"
    xmlns:f="jakarta.faces.core"
    xmlns:ui="jakarta.faces.facelets"
    xmlns:pt="jakarta.faces.passthrough">
<ui:composition>

<h:inputText id="#{idParam}" value="#{valueParam}">
  <f:validateRegex pattern="\d{4}-\d{2}-\d{2}" />
</h:inputText>

</ui:composition>
</html>
```

Remember that the code around the <ui:composition> tag will be disregarded, so you add <html> to declare the namespaces and make this a well-formed XHTML document. The #{idParam} and #{valueParam} tags are the component parameters.

Next, create a file called mytaglib.taglib.xml inside src/main/webapp/WEB-INF. For its contents, write the following:

```
<?xml version="1.0" encoding="UTF-8"?>
<facelet-taglib xmlns="https://jakarta.ee/xml/ns/jakartaee"
  xmlns:xsi="http://www.w3.org/2001/XMLSchema-instance"
  xsi:schemaLocation="https://jakarta.ee/xml/ns/jakartaee
  https://jakarta.ee/xml/ns/jakartaee/
  web-facelettaglibary_4_0.xsd" version="4.0">
    <namespace>
      http://book.jakartapro.mytaglib
    </namespace>
    <tag>
        <tag-name>fancyDate</tag-name>
        <source>tags/fancyDateInput.xhtml</source>
```

```
    </tag>
</facelet-taglib>
```

(Remove the line break and spaces after jakartaee/.) The namespace can be anything, but it must be unique and should certainly point to your company.

Now you must register the new tag library. To this aim, add the following to the src/main/webapp/WEB-INF/web.xml file:

```
<?xml version="1.0" encoding="UTF-8"?>
<web-app ...>
  ...
  <context-param>
    <param-name>javax.faces.FACELETS_LIBRARIES</param-name>
    <param-value>/WEB-INF/mytaglib.taglib.xml</param-value>
  </context-param>
  ...
</web-app>
```

To use that new component in any XHTML page, you first have to declare the new, custom namespace. After that, you can use it like any other component:

```
<!DOCTYPE html>
<html xmlns:h="jakarta.faces.html"
    xmlns:f="jakarta.faces.core"
    xmlns:ui="jakarta.faces.facelets"
    xmlns:pt="jakarta.faces.passthrough"
    xmlns:mytags="http://book.jakartapro.mytaglib">
<h:body>
    ...
    <mytags:fancyDate idParam="theDate3"
        valueParam="#{someBean.someField}" />
    ...
</h:body>
</html>
```

You can see that the attribute names refer to the #- parameters used in the fancyDateInput.xhtml file.

Composite Components

Composite components are more lightweight compared to custom tag libraries. While they implicitly define a new non-public namespace, you don't have to define and register a tag library in order to use them. The steps to define a composite component are as follows:

1. Make up a component collection name. This could be anything. This example uses mycomponents.

2. Create a file called myComponent.xhtml. You can use any name here, but you must jot down the filename (without the .xhtml), because it will be the name of the component. Save the file inside src/main/webapp/resources/mycomponents.

3. Use <composite:interface> and <composite:implementation> to define an interface and an implementation for the new component. See Figure 8-1.

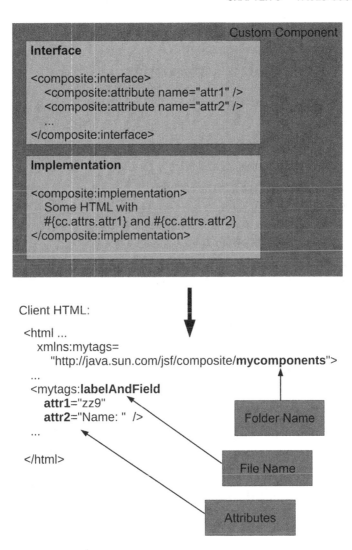

Figure 8-1. *Custom components*

The new component can then be used from any Faces file by using the http://java. su.com/Faces/composite/mycomponents namespace and adding attributes with names defined in the interface.

As an example, consider the labelAndField component, which combines a label and a text input field. Create a file called labelAndField.xhtml and put it in the src/ main/webapp/resources/mycomponents folder:

```
<html xmlns:h="jakarta.faces.html"
  xmlns:f="jakarta.faces.core"
```

```
  xmlns:ui="jakarta.faces.facelets"
  xmlns:pt="jakarta.faces.passthrough"
  xmlns:composite="jakarta.faces.composite">

<!-- ............... INTERFACE ............... -->
<composite:interface>
    <composite:attribute name="fieldId" />
    <composite:attribute name="labelTxt" />
    <composite:attribute name="bindTo" />
</composite:interface>

<!-- ............... IMPLEMENTATION ........... -->
<composite:implementation>
  <h:outputLabel for="#{cc.attrs.fieldId}"
      value="#{cc.attrs.labelTxt}" />
  <h:inputText id="#{cc.attrs.fieldId}"
      value="#{cc.attrs.bindTo}" />
</composite:implementation>

</html>
```

Observe the new namespace declaration called jakarta.faces.composite. In the implementation, you refer to parameters via #{cc.attrs.ATTRNAME}. This is a convention; you *have to* use cc.attrs. as a prefix.

To use the new component, declare the composite namespace with mycomponents added:

```
<!DOCTYPE html>
<html xmlns:h="jakarta.faces.html"
  xmlns:f="jakarta.faces.core"
  xmlns:ui="jakarta.faces.facelets"
  xmlns:pt="jakarta.faces.passthrough"
  xmlns:mytags="jakarta.faces.composite/mycomponents">
<h:body>
  ...
  <mytags:labelAndField
      fieldId="zz9"
      labelTxt="The Label: "
```

```
        bindTo="#{someBean.someField}" />
  ...
</h:body>
</html>
```

Custom Components in Java

Although it's a highly advanced operation, it is possible to define your own components with a Java class. This might be necessary if component programming using just XHTML files doesn't suit your needs.

To create such a class, add the appropriate annotations and let the class extend UIComponent, as follows:

```
package this.package;

import jakarta.faces.component.FacesComponent;
import jakarta.faces.component.UIComponent;

@FacesComponent(createTag = true,
        namespace = "https://mycompany.com/mynamespace",
        tagName = "custom-component",
        value = "this.package.CustomComponent")
public class CustomInput extends UIComponent {

    ...

}
```

If you do it this way, you have to implement a lot of abstract methods from UIComponent. It is maybe a little bit easier if you extend from UIInput or UIOutput, or at least from UIComponentBase. The most important methods to override are the decode... () and encode...() methods, since those define the rendering and user input value extraction procedures of the component.

The API documentation of UIComponent and its derived classes give you more information about such Faces Java components. It also helps to look at the implementations of the standard components included with Faces.

Once you have properly implemented such a component, it is easy to use. Just write the following in any Faces page:

```
<!DOCTYPE html>
<html xmlns:h="jakarta.faces.html"
  xmlns:f="jakarta.faces.core"
  xmlns:ui="jakarta.faces.facelets"
  xmlns:pt="jakarta.faces.passthrough"
  xmlns:mytags="https://mycompany.com/mynamespace">
<h:body>
  ...
  <mytags:custom-component ...>
      ...
  </mytags:custom-component>
  ...
</h:body>
</html>
```

CHAPTER 9

Flows

Faces Flows is a page-cohesion concept that entered the JEE world in 2013 with JEE version 7. Its main idea is that a couple of pages are said to belong to a "flow." Data associated with such flows and stored on the server can be confined to a *flow scope*, which means that all pages of a flow can access the same common flow scope data pool, and that the associated memory can be freed once the flow is exited. A flow variable scope thus lives between the *request scope,* where variables only survive a request, and the *session scope,* where data lives as long as a user session prevails.

The predominant usage scenario for flows is a wizard, where the pages belong to a chain of interconnected dialog-like input pages and serve a precise use case, like a payment process, a poll, a registration workflow, or similar.

The Flow Process

To define a flow, you usually create a couple of XHTML pages inside a dedicated folder, called `src/main/webapp/flow-name`. Instead of `flow-name`, you can use any suitable name. To then enter the flow, you can use a button inside your web application, like so:

```
<h:commandButton value="Enter Flow"
     action="flow-name"/>
```

To exit the flow, you provide a file called `flow-name-return.xhtml` *outside* the flow folder and write anywhere in the flow XHTML pages:

```
<h:commandButton value="Finish"
     action="flow-name-return" />
```

Replace `flow-name` with the name you actually chose for the flow. For the data handed over between the flow pages, you provide a bean annotated with `@FlowScoped( value = "flow-name" )`, again using your flow name for the attribute value. See Figure 9-1.

© Peter Späth 2023
P. Späth, *Pro Jakarta EE 10*, https://doi.org/10.1007/978-1-4842-8214-4_9

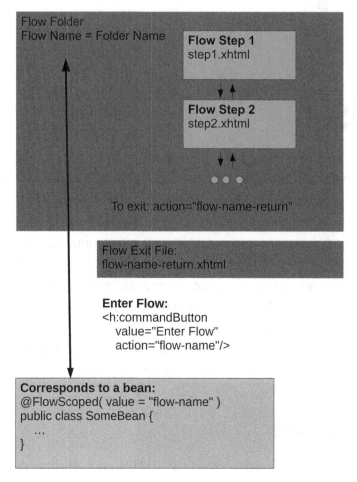

Figure 9-1. *The flow process*

Basic Flow Setup

Flows follow the convention-over-configuration paradigm. You therefore don't need a configuration for a basic flow. The convention is as follows:

1. Create a folder called `flow-name` inside `src/main/webapp`. You can choose any folder name you like, but by convention the folder name will also be the flow name.

2. Create an empty file called flow-name-flow.xml inside src/main/
   webapp/flow-name. The first part of the filename, flow-name, *must*
   match the flow's name. So if your flow is called flow-a instead,
   you must have a folder called src/main/webapp/flow-a and this
   empty file must have the name flow-a-flow.xml. Because this file
   exists, Faces knows that this folder belongs to a flow. If you later
   need to configure the flow, the configuration will go into this file.

3. Create a flow start page called flow-name.xhtml inside
   src/main/webapp/flow-name. The filename without the suffix,
   flow-name, *must* match the flow's name. This is a standard Faces
   XHTML page.

4. Create any number of subsequent flow pages inside src/main/
   webapp/flow-name. There is no restriction to the names to
   be used.

5. To save data (user input) in the flow and handle navigation cases,
   create a bean class with annotations @Named (import "jakarta.
   inject") and @FlowScoped(value = "flow-name"). This bean
   must also implement java.io.Serializable. This is a normal
   backing bean, but it is only available for XHTML pages inside the
   flow. The value attribute *must* match the flow name.

6. Create a file named flow-name-return.xhtml *outside* the
   src/main/webapp/flow-name folder. The first part of the filename,
   everything in front of the -return.xhtml, *must* match the flow's
   name. This is the exit point of the flow. It is a standard Faces
   XHTML page, but you cannot use the @FlowScoped bean there—it
   will have been disposed of before the page gets called.

See Figure 9-2 (the flow has the name flow-a in this figure).

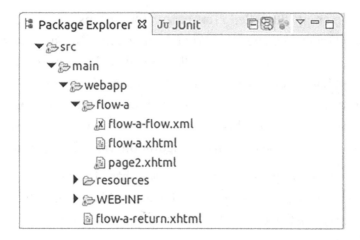

Figure 9-2. *A basic flow setup*

To enter the flow, you use the folder name, as in the following:

```
...
<h:form>
  ...
  <h:commandButton value="Enter Flow"
        action="flow-name"/>
</h:form>
```

To exit the last flow page, use the exit page name without a path or suffix, as follows:

```
...
<h:form>
  ...
  <h:commandButton value="Finish"
        action="flow-name-return" />
</h:form>
```

Where you substitute the actual flow name plus -return for flow-name-return if you changed this.

A basic flow scoped bean with the necessary annotations and an example field reads as follows:

```
package the.package.name;

import java.io.Serializable;

import jakarta.faces.flow.FlowScoped;
import jakarta.inject.Named;

@Named
@FlowScoped(value = "flow-name")
public class FlowScopeBean implements Serializable{
    private static final long serialVersionUID =
        -3877099938046102267L;

    private String value;

    public String getValue() {
      return value;
    }

    public void setValue(String value) {
      this.value = value;
    }
}
```

Overriding Conventions

If you don't want to use the conventions you just read about, you must provide a suitable configuration to Faces. One possibility is to add configuration entries inside this file:

```
src/main/webapp/flow-name/flow-name-flow.xml
```

Where flow-name in both cases stands for the flow name, so you must replace both if your flow has a different name.

The contents of this file would read as follows:

```
<faces-config
    xmlns="https://jakarta.ee/xml/ns/jakartaee"
    xmlns:xsi="http://www.w3.org/2001/XMLSchema-instance"
    xsi:schemaLocation="https://jakarta.ee/xml/ns/jakartaee
```

```
        https://jakarta.ee/xml/ns/jakartaee/
        web-facesconfig_4_0.xsd"
    version="4.0">

    <flow-definition id="flow-name">
        ...
    </flow-definition>
</faces-config>
```

(Remove the line break and the spaces after `jakartaee/`). The `id` attribute must match the folder name used for the flow.

Specifying a Different Flow Start Page

To specify a different flow start page, write the following inside the `<flow-definition>` element:

```
<flow-definition id="flow-name">
    ...
    <view id="flow-a-start">
      <vdl-document>/flow-name/page1.xhtml</vdl-document>
    </view>
    <start-node>flow-a-start</start-node>

</flow-definition>
```

For the ID attribute inside `<view>`, you can choose whatever you like, but it must match the contents of the `<start-node>` element.

Specifying a Different Return Page

If you want to use a different return page for the flow, write the following inside the `<flow-definition>` element:

```
<flow-definition id="flow-name">
    ...
    <flow-return id="returnFromFlow">
      <!-- <from-outcome>#{someBean.returnValue}</from-outcome>  -->
```

```
<from-outcome>/start</from-outcome>
  </flow-return>

</flow-definition>
```

For the ID attribute inside `<flow-return>`, use the same value as for the finishing command button in the last flow page:

```
...
<h:form>
  ...
  <h:commandButton value="Finish"
        action="returnFromFlow" />
</h:form>
```

The `<from-outcome> /start </from-outcome>` will go to `start.xhtml`. If you use a method expression instead, as in `<from-outcome> #{someBean.aboutToExit}` `</from-outcome>`, this method will be called instead to determine where to go:

```
public String aboutToExit() {
  return "start"; // -> start.xhtml
}
```

Programmatic Configuration

Flows can also be configured programmatically using a Java class. In order for Faces to recognize the flow configuration, this class' name must match the flow name. So, if you for example have a flow folder called checkout, the name of this configuration class must be CheckoutFlow (first letter must be uppercase, with Flow added).

To specify a custom flow start page and a custom flow return page, this class reads:

```
package the.package;

import java.io.Serializable;

import jakarta.enterprise.inject.Produces;
import jakarta.faces.flow.Flow;
import jakarta.faces.flow.builder.FlowBuilder;
import jakarta.faces.flow.builder.FlowBuilderParameter;
```

```
import jakarta.faces.flow.builder.FlowDefinition;

// The name implies that the flow files are inside
// folder "flow-a"
public class FlowAFlow implements Serializable {
  private static final long serialVersionUID = 1586568679L;

  @Produces
  @FlowDefinition
  public Flow defineFlow(@FlowBuilderParameter FlowBuilder
        flowBuilder) {

    // Specify a flow start page
    String flowId = "flow-a";
    flowBuilder.id("", flowId);
    flowBuilder.viewNode(flowId, "/" + flowId +
        "/flow-a-start.xhtml").markAsStartNode();

    // Specify a flow return page
    flowBuilder.returnNode("returnFromFlow").
        fromOutcome("/start");

    return flowBuilder.getFlow();
  }
}
```

In .fromOutcome() you can also specify a method expression:

```
...
.fromOutcome("#{someBean.aboutToExit}");
```

Note   Be aware that, by using EL constructs inside Java classes, you introduce a two-fold structuring of class dependencies: Java class relations and EL relations.

Handling Flow Outcome

Before you return to the exit page called `flow-name-return.xhtml`, you can invoke one last navigation case inside the `@FlowScope`'d backing bean. You can, for example, evaluate all data entered by the user during the flow and perform the flow's actual purpose. To this aim, you can write a Finish button as follows:

```
...
<h:form>
    ...
    <h:commandButton value="Finish"
        action="#{flowScopeBean.aboutToExit}" />
</h:form>
...
```

In the bean, you then add a method to perform any finishing actions:

```
public String aboutToExit() {
  // Here we can perform activities with the data
  // entered during the flow...
  return "flow-name-return"; // flow name + "-return"
}
```

Passing Data Between Flows

Flows can invoke other flows. However, this cannot be done by simply using the second flow's name in a `<h:commandButton>` component. Instead, you must explicitly declare that flow transition in the first flow's configuration.

If you want to do this in the XML configuration, add the following to the `flow-name/flow-nameflow.xml` file:

```
<flow-definition id="flow-name">
  ...
  <!-- possibly: -->
  <flow-return ...>
      ....
  </flow-return>
```

```
<flow-call id="call-flow-b">
    <flow-reference>
        <flow-id>flow-b</flow-id>
    </flow-reference>
</flow-call>
</flow-definition>
```

where the `call-flow-b` ID in `<flow-call>` corresponds to a command button action somewhere in the calling flow:

```
<h:commandButton value="Flow B" action="call-flow-b" />
```

The `<flow-reference><flow-id>...</flow-id></flow-reference>` corresponds to the second flow called.

You cannot use flow-scoped variables to pass data from one flow to another. Instead, you add an *outbound* parameter specification in the configuration of the calling flow:

```
<flow-call id="call-flow-b">
  <flow-reference>
    <flow-id>flow-b</flow-id>
  </flow-reference>
  <outbound-parameter>
    <name>param1FromCallingFlow</name>
    <value>the value</value>
  </outbound-parameter>
</flow-call>
```

where instead of `the value`, you can also use an EL value expression.

Inside the called flow's configuration, you then add an *inbound* parameter specification:

```
<flow-definition id="flow-b">
  <inbound-parameter>
    <name>param1FromCallingFlow</name>
    <value>#{someFlowBean.value}</value>
  </inbound-parameter>
</flow-definition>
```

where the `someFlow` bean is a backing bean with flow scope *inside* the called flow.

> **Caution**   Again, because of a bug in the schema definition file, these
> configuration entries will be marked as erroneous. However, the configuration
> shown here was tested with GlassFish 7, where it worked without issues.

The same configuration using Java classes instead of XML files can be achieved using the following:

```
// inside the calling flow:
flowBuilder.flowCallNode("call-flow-b").
        flowReference("", "flow-b").
        outboundParameter("param1FromCallingFlow",
            "the value");

// inside the called flow:
flowBuilder.inboundParameter("param1FromCallingFlow",
            "#{someFlowBean.value}");
```

## CHAPTER 10

Websockets

HTTP by its nature is a one-directional, stateless protocol. This means one party, a client like a web browser, sends a single request to the server and waits for the returned data, eventually closing the connection once all the data from that single request is retrieved. From the protocol perspective, any subsequent request does not know about previous requests, which is why HTTP is called a stateless protocol.

However, applications on a server need to maintain session context once it is necessary to assign several HTTP requests to a particular user operating the application. This was made possible by attaching session information as GET parameters or by using cookies. Thus HTTP remains stateless and adding state does not contradict HTTP's nature.

The conceptional unidirectional nature of HTTP is a different story. While at the beginning of the Internet it was completely acceptable that a data flow could only be initiated by the client, with modern web applications, it became more and more necessary that data flows could be triggered by the server as well. Think of a chat application where the server needs to inform user B that user A has sent some message. Several techniques have been used since then to overcome this conceptional deficiency of HTTP, but neither of them was easy to understand and/or to maintain.

As a remedy, *websockets* were invented, which allow bidirectional communication between servers and clients. Since JEE version 7, websockets were included with Java enterprise servers. Jakarta EE 10 of course includes a websockets implementation as well. This chapter talks about how to use websockets with Jakarta EE.

Websockets on the Server Side

Websockets are covered by JSR 356. You can view the specification at

```
https://www.jcp.org/en/jsr/detail?id=356
```

© Peter Späth 2023
P. Späth, *Pro Jakarta EE 10*, https://doi.org/10.1007/978-1-4842-8214-4_10

There are several environments where server-side websockets can be run. Since this book talks about Jakarta EE, this chapter consider websockets in a web application bundled by a WAR as the predominant place to run them.

You start a websocket project like any other WAR project described in this book. To add a websocket, all you have to do is introduce an appropriately annotated Java class. Choose any name and package and write the following:

```
package book.jakartapro.websocket.server;

import jakarta.enterprise.context.ApplicationScoped;
import jakarta.inject.Inject;
import jakarta.json.Json;
import jakarta.json.JsonObject;
import jakarta.json.JsonReader;
import jakarta.websocket.OnClose;
import jakarta.websocket.OnError;
import jakarta.websocket.OnMessage;
import jakarta.websocket.OnOpen;
import jakarta.websocket.Session;
import jakarta.websocket.server.ServerEndpoint;

@ApplicationScoped
@ServerEndpoint("/actions")
public class WebSocketServer {
    @OnOpen
    public void open(Session session) {
        // Somehow handle an opened session. For example
        // save the session object in the instance of
        // some session registrar class...
    }

    @OnClose
    public void close(Session session) {
        // Somehow handle a closed session
    }

    @OnError
```

```java
public void onError(Throwable error) {
    // Somehow handle a websocket error
}

@OnMessage
public void handleMessage(String message,
        Session session) {
    // Receive a string message. It is not required
    // to use JSON as a message format, but it is a
    // good choice in many cases and it simplifies
    // programming JavaScript websocket clients
    try (JsonReader reader = Json.createReader(
            new StringReader(message))) {
        JsonObject jsonMessage = reader.readObject();

        // handle the incoming message...
    }
}
}
```

You can see that you added the server object to the application scope. The usual session scope won't work here, since websockets do not correspond to user sessions in the traditional HTTP sense.

In order to send messages to a client, you get a hold of a `jakarta.websocket.Session` object and then issue messages as follows:

```java
// Build a data object:
JsonProvider provider = JsonProvider.provider();
JsonObject msg = provider.createObjectBuilder()
    .add("someField", "someValue")
    .add("someOtherField", 42)
    ...
    .build();

// Send it to the client:
try {
    session.getBasicRemote().sendText(msg.toString());
} catch (IOException ex) {
```

```
    // Handle errors. Also maybe remove the session
    // from the session storage:
    // sessions.remove(session);
}
```

If the WAR file runs under the `WebSock-Server-War` context, the websocket URI on a local GlassFish reads as follows:

```
http://localhost:8080/WebSock-Server-War/actions
```

You need to know that in order for a client web application (JavaScript) to connect to the websocket.

By the way, in order to change the context of the EAR plugin (in case the web application is part of an EAR), you must adapt the EAR plugin configuration inside `build.gradle`:

```
...
ear {
    deploymentDescriptor {
        webModule("WebSock-Server-War.war",
                  "sock-context")
    }
}
```

where `WebSock-Server-War.war` is the generated WAR module's deployment artifact and `sock-context` is the new context:

```
http://localhost:8080/sock-context/
```

If you instead deploy the web application as a WAR file, it is the server's business to determine how the context looks. For GlassFish 7, you can specify the context root during the deployment:

```
./asadmin \
    --user THE_USER \
    deploy --contextroot sock-context theWar.war
```

Websockets on the Client Side

On the client side, you add the following JavaScript code to your web application:

```
<!DOCTYPE html>
<html>
<head>
<title></title>
<meta http-equiv="Content-Type"
    content="text/html; charset=UTF-8">
<script>
  var server = "localhost:8080/WebSock-Server-War";
  var socket = new WebSocket("ws://" + server +
        "/actions");
  socket.onmessage = onMessage;

  function onMessage(event) {
    // Handle messages which come from the server...
    // We assume the message is in JSON format
    var msg = JSON.parse(event.data);
    ...
  }

  ...

  // Sending a message to the server. You do this from
  // somewhere in your JavaScript code or an
  // event handler like a button press
  var msg = {
      id: 42,
      txt: "Hello World"
  };
  sendToServer(msg);

  ...

  function sendToServer(msg) {
    // We convert the message to a JSON string, so
```

```
    // 'msg' could be any object
    socket.send(JSON.stringify(msg));
  }
</script>
<link rel="stylesheet" type="text/css"
      href="css/style.css">
</head>
<body>
  ...
</body>
</html>
```

The /actions part from the socket URL should match the @ServerEndpoint annotation from the Java server class.

CHAPTER 11

Frontend Technologies

With web applications situated on the interface between users and application functionality, and with the need to present well designed and highly usable frontends on various types of devices, frontend technologies have gained lots of attention in the developer community.

It is acceptable to build web frontends using the dedicated Jakarta EE technologies Faces or REST, but with the plethora of different and sometimes even contradictory requirements that a frontend has to fulfill, there are many different web frontend technologies. It is not easy to decide whether you need a third-party web frontend technology, and, if you do, which one to choose. These basic questions can help you make a decision:

- Does it cost something, or is it open-source or free?

- How good is the documentation?

- What is the expected effort to produce a good design?

- How easy is it to merge graphical designers' ideas into the web application?

- How easy is it to ensure the web application has good usability?

- How difficult is it to include the frontend in Jakarta EE?

- How easy is it to address different frontend devices (browsers and smartphones)?

- How easy is it to get help if problems arise?

- How easy is it to upgrade to new versions?

- Which programming skills are required to use the frontend technology?

- How likely is it that the technology will still exist in five years?

© Peter Späth 2023
P. Späth, *Pro Jakarta EE 10*, https://doi.org/10.1007/978-1-4842-8214-4_11

From the number of questions and possible answers, you might guess that it is easy to list ten, twenty, or more frontend technologies, all of them with merits and drawbacks. It is impossible to provide a thorough introduction to all of them in a single book, but the following sections give an overview of several and present some pros and cons.

Note   Due to the nature of such comparisons, it is possible that advocates of one or the other framework will disagree with my assessments. Note, however, that this is only a rough comparison of features. You can use each product's web documentation to draw your own conclusions.

Walking through the list, you might get the impression that I'm not a fan of Node.js. I have to admit that this is indeed the case. I don't follow the usual argument that web developers know how to use JavaScript and so using JavaScript as a server-side language and build environment is a good idea. Instead, I think that the Java ecosystem already has everything that is needed to build good and elegant code, and that the frontend logic running in a browser and the backend logic should stay separate, for both have different requirements concerning architectural aspects and developer programming skills. That said, it is of course up to you. Go ahead and use Node.js to any extent you like. I personally think that mixing Jakarta EE and Node.js introduces unnecessary conceptional complexity in a project process.

No Third-Party Frontend Technology

As mentioned, the Faces and REST technologies included in Jakarta EE are enough to build any kind of web application. You can simplify development if you include the jQuery JavaScript library, because it evens out some browser differences and helps you write shorter, more elegant JavaScript code.

Pros:

- Faces and REST are already included in Jakarta EE

- Heavily standardized and stable

- Very good documentation

Cons:

- Advanced development skills needed to achieve a good graphical design and high usability

- Different frontend devices are not explicitly addressed

Play 2

Play (version 2) is a full-stack web application technology. It works with Java or Scala as a programming language and allows you to use functional patterns. It follows the MVC (model view controller) paradigm. A basic development project including a running web server is set up in five minutes.

```
https://www.playframework.com/
```

Pros:

- HTTP based, stateless.

- Explicitly addresses mobile devices.

- Asynchronous requests, non-blocking I/O.

- Rapid development lifecycle.

Cons:

- Some bias toward Scala instead of Java as a programming language.

- While Play supports both SBT and Gradle as a build tool, the introduction basically relies on SBT, which requires you to learn SBT to get acquainted with Play.

- Play introduces a full-stack web application technology. It is therefore a competitor to Jakarta EE and mixing both is challenging.

React

React is a component-based web development framework. You use an augmented JavaScript language named JSX and mix programming logic and HTML markup. For example:

```
class ShoppingList extends React.Component {
  render() {
    return (
      <div className="shopping-list">
        <h1>Shopping List for {this.props.name}</h1>
        <ul>
          <li>Bananas</li>
          <li>Apples</li>
          <li>Peanuts</li>
        </ul>
      </div>
    );
  }
}
```

An on-the-fly compiler transforms JSX into JavaScript during development, and for production, a pre-compiler is available.

```
https://reactjs.org/
```

Pros:

- Mixes JavaScript and HTML coding (JSX pages), somewhat simplifying development.

- Seamlessly integrates with other technologies, so you can include React features in Jakarta EE applications step-by-step.

- Development goes with an on-the-fly compilation of JSX pages, which makes development fast.

Cons:

- React is maintained by Facebook, which is dependent on company policies. You might not want to accept these preferences for your software. Also, the license includes some parts that feel like vendor lock-in.

- The production code build process heavily depends on Node.js, which introduces a new paradigm based on JavaScript as a server code language. Using React means you have to maintain two ecosystems, Java and JavaScript. Also, Node.js introduces another vendor lock-in: Private and company owned software repositories for Node.js modules are not easy to build and maintain.

- The markup (HTML) and logic are not separated, which contradicts the widely accepted MVC paradigm.

Angular 2

Angular is a templating web development framework. You create HTML pages with placeholders for Angular components in a special syntax. Deploying Angular applications happens via Node.js (a server-side JavaScript framework).

```
https://angular.io/
```

Pros:

- Explicitly addresses mobile devices.

- Angular transports *dependency injection* into the web layer (JavaScript/TypeScript code).

Cons:

- Angular depends on TypeScript, which needs to be transformed into JavaScript before it can run on a browser.

- Angular heavily depends on Node.js, which introduces a new paradigm based on JavaScript as a server code language. Using Angular with Jakarta EE means you have to maintain two ecosystems—Java and JavaScript. Also, Node.js introduces another vendor lock-in: Private and company owned software repositories for Node.js modules are not easy to build and maintain.

- Angular introduces a full-stack web application technology. It is therefore a competitor to Jakarta EE and mixing both in a stable manner is challenging.

Spring Boot

Spring Boot utilizes the Spring platform to allow you to rapidly develop web applications.

```
https://spring.io/projects/spring-boot
```

Pros:

- Quick setup of web applications (and microservices) in a standalone, almost production-ready environment.

- Spring Boot automatically opens monitoring ports for discovering metrics and technical application status.

- Heavily relies on the *convention over configuration* pattern, simplifying configuration and speeding up development.

Cons:

- Spring might be considered a competitor to Jakarta EE, although Spring and Jakarta mutually adopt features from each other during their version upgrades. Mixing both is possible, but doing so introduces a new level of complexity to the application development.

- Spring Boot is a "Let Others Do Your Job" framework. You somehow lose control over the internal function of your application. This is something you might not want to do.

- Spring Boot has a bias toward distributed microservices. If this contradicts your application's architecture, Spring Boot might not be your best choice.

Vue

Vue is a templating framework, similar to Angular, but with less complexity and a more modular approach.

```
https://vuejs.org
```

Pros:

- Uses JavaScript as a frontend logic idiom, so there is no need to learn TypeScript as is the case for many other frameworks.

- Vue is incrementally adoptable, meaning you can introduce Vue step-by-step in your web application.

- Allows for fast development, since all templating happens on-the-fly by the pages' JavaScript code.

- While being optional, it is possible to render HTML pages on the server, speeding up page loading (at the cost of server load, of course).

Cons:

- While Vue can be included in your web application by simply using a `<script>` tag in the web applications HTML page(s), the documentation suggests using Node.js for a build environment. This introduces a new paradigm based on JavaScript as a script language.

Spring MVC

Spring MVC is the less "convention over configuration" sibling of Spring Boot. What you read about Spring Boot previously holds for Spring MVC, except that Spring MVC requires more configuration work. This, in turn, leads to an increased versatility of Spring MVC if compared to Spring Boot, with the cost of more work to be done to get it working.

```
https://spring.io/guides/gs/serving-web-content/
```

Pros:

- Lines up with the Spring architecture, with a component hierarchy built up by injection.

- Good documentation.

Cons:

- High grade of complexity.

- Spring might be considered a competitor to Jakarta EE, although Spring and Jakarta mutually adopt features from each other during their version upgrades. Mixing both is possible, but introduces a new level of complexity to the application development.

- Spring is a "let others do your job" framework. By using Spring sub-technologies or Spring modules, you somehow lose control over the internal function of your application. This is something you might not want to do.

Ember

Ember is another template-based web application framework that relies on Node.js as a build and execution framework.

```
https://emberjs.com/
```

Pros:

- Heavily *convention over configurations* based, allowing for Lean code.
Cons:

- Although *convention over configuration* based, Ember seemingly has a steeper learning curve compared to other frameworks.

- Ember depends on Node.js, which introduces a new paradigm based on JavaScript as a server code language. Using Ember together with Jakarta EE means you have to maintain two ecosystems—Java and JavaScript. Also, Node.js introduces another vendor lock-in: Private and company owned software repositories for Node.js modules are not easy to build and maintain.

- Ember introduces a full-stack web application technology. It is therefore a competitor to Jakarta EE and mixing both in a stable manner is challenging.

Act.Framework

```
http://actframework.org/
```

Pros:

- Uses dependency injection, allowing for Lean code.

- By relying on REST for frontend-to-backend communication, exhibits a clear and high-performing architecture.

- Uses Maven for builds, therefore better integrates with the Jakarta EE build processes.

- Allows you to choose between different rendering (templating) engines.

Cons:

- Templating is done by third-party libraries, which leads to leaving the Jakarta EE specification. While code might be smaller compared to Faces, using `Act.Framework` introduces additional project complexity due to the need to monitor backend (Jakarta EE) and frontend (`Act.Framework`) specification development.

Apache Struts 2

Apache Struts is a venerable frontend technology that follows the MVC (model view controller) paradigm.

```
https://struts.apache.org
```

Pros:

- By design, Apache Struts tries not to hide the stateless nature of the HTTP protocol. Thus Struts might be considered a "clean" web development approach.

- Struts is a seasoned web development framework showing a high grade of stability and is expected to be future-proof.

Cons:

- There is a steep learning curve and some proficiency in understanding the HTTP request-response cycle is required.

- While Struts 2 supports several frontend templating engines or can even go without elaborate templating by utilizing REST, Struts exhibits some bias toward the "old fashioned" JSP technology.

GWT

GWT by Google Inc. is basically a compilation engine transforming Java classes into JavaScript, which then can be executed in a browser.

```
http://www.gwtproject.org/
```

Pros:

- Takes away the need to develop any JavaScript code. Everything is just Java.

- It is possible to migrate from a common web application to GWT iteratively.

- Good integration into IDEs like Eclipse.

- Good debugging support.

Cons:

- You are somewhat limited to build web applications the way the Google people think of.

- Java by itself is not good at defining HTML markup. The transition from Java to JavaScript and HTML, although automated by GWT, is far away from being a simple, clean procedure.

- There is no easy way to do things. A first working web application version will show up much later compared to other web application frameworks.

- Writing good GWT applications requires a lot of Java development skills.

Vaadin

Vaadin is a Java-based web development framework. The distribution distinguishes between a free variant with basic features and a paid variant with more elaborated modules.

```
https://vaadin.com/
```

Pros:

- Explicitly addresses PC browsers and handhelds.

- Includes elaborated components for grid data, charts, and dialogs.

- Professional looking UI design by default.

- Explicit Java EE (/Jakarta EE) support.

- Maven and Gradle build tooling.

- You only need Java for the complete web application.

Cons:

- Separated into free basic features and paid pro features. This always bears the risk that one day you might be forced to pay for features you formerly considered as basic.

- The UI logic and state is stored on the server. This leads to a performance impact for heavy-load publicly available applications.

- Fine-grained frontend customization, for example animation, is hard to achieve.

DataTables

This is a `jQuery` plugin. While it's not a multi-purpose frontend technology, `DataTables` enables you to show and edit tabular data on the browser in a very versatile manner. In in-house applications, listing tabular data and editing the data are common tasks, and `DataTables` helps you rapidly implement such applications with Jakarta EE running in the background.

```
https://www.datatables.net/
```

Pros:

- Easy integration. As a jQuery plugin, `DataTables` imposes no restrictions on the backend technology in use.

- Highly flexible.

- Powerful search features.

- Well documented.

Cons:

- Not a multi-purpose frontend technology.

D3js

D3js is a powerful visualization library. It helps you visualize scientific and/or business related data.

`https://d3js.org/`

Pros:

- Very flexible visualization library.

- Highly professional looking design.

- Simple inclusion by HTML `<script>` tags.

Cons:

- Not a multi-purpose frontend technology.

- It takes some time to get an overview of all the features

- To use it, good JavaScript skills are required.

## CHAPTER 12

Form-Based Authentication

Secured web applications typically use the HTTPS (SSL/TLS) communication protocol rather than HTTP and require users to authenticate themselves via a login dialog box. With the *basic authentication* procedure, the browser presents a standardized input form for the user credentials. Although it's easy enough to implement and configure, the basic authentication does not allow you to control the presentation and behavior of the login dialog box. Professional web applications use *form-based authentication* instead, which allows them to implement their own login dialog box that's compliant with company policies and design guidelines.

Note JSR-375 attempted to standardize the Jakarta EE security techniques. However, the implementation seems to be in a questionable state, apart from announcements that you cannot find helpful tutorials and examples on the Internet. I therefore describe the somewhat old fashioned but reliably working and well documented original Jakarta EE security enhancements techniques in this book.

Enabling Security on the Sever

Enabling and configuring security on the server, as well as adding users and groups, depends on the Jakarta EE server product, its configuration, and the security methods you use. Although this means that you have to read the server documentation for specifics, the following paragraphs describe the basic steps for GlassFish 7 to give you a basic idea.

© Peter Späth 2023
P. Späth, *Pro Jakarta EE 10*, https://doi.org/10.1007/978-1-4842-8214-4_12

Note   We are not dealing with users stored in some dedicated SQL user database. To include database access, create a new realm of type `com.sun.enterprise.security.auth.realm.jdbc.JDBCRealm`. Using the terminal, you can enter the following on one line:

```
asadmin create-auth-realm
    --classname com.sun.enterprise.security.auth.realm.jdbc.JDBCRealm
    --property jaas-context=jdbcRealm
      :datasource-jndi="jdbc/auth"
      :user-table=users
      :user-name-column=email
      :password-column=password
      :group-table=users_groups
      :group-name-column=groupname
      :digest-algorithm=SHA-512
  userMgmtJdbcRealm
```

If you want to use the web admin console, you must add the properties one by one while creating the realm.

First, you have to enable security in a server-wide configuration setting. To achieve this, in the web administration console at `http://localhost:4848`, choose Configurations ➤ Server-config ➤ Security. Then check the Security Manager checkbox.

Next, you need to add a sample user called `AdminUser` to the server's user database. The user has to be assigned to the `ApplAdmin` group. Again in `http://localhost:4848`, navigate to Configurations ➤ Server-config ➤ Security ➤ Realms ➤ File. Click the Manage Users button. On the page that appears, click New and then enter the following:

```
User ID:    AdminUser
Group List: ApplAdmin
Password:   pW41834
```

Click OK. This new user is stored in a file inside the server's file-tree structure. This is where the name `file` realm comes from.

Form-Based Authentication for Faces

To start developing form-based authentication for your Faces pages, create a new Gradle project, and as usual, add a `src/main/webapp/WEB-INF` folder containing four files. A file called `beans.xml`:

```xml
<?xml version="1.0" encoding="UTF-8"?>
<beans xmlns="https://jakarta.ee/xml/ns/jakartaee"
    xmlns:xsi="http://www.w3.org/2001/XMLSchema-instance"
    xsi:schemaLocation="https://jakarta.ee/xml/ns/jakartaee
        https://jakarta.ee/xml/ns/jakartaee/beans_3_0.xsd"
    bean-discovery-mode="all" version="3.0">
</beans>
```

And three more files, called `faces-config.xml`, `glassfish-web.xml`, and `web.xml`. For the contents of `faces-config.xml`, write the following:

```xml
<?xml version="1.0" encoding="UTF-8"?>
<faces-config
  xmlns=
    "http://xmlns.jcp.org/xml/ns/javaee"
  xmlns:xsi=
    "http://www.w3.org/2001/XMLSchema-instance"
  xsi:schemaLocation=
    "http://xmlns.jcp.org/xml/ns/javaee
    http://xmlns.jcp.org/xml/ns/javaee/
        web-facesconfig_2_3.xsd"
  version="2.3">
  <application>
    <resource-bundle>
      <base-name>
        book.jakartapro.formbasedauth.web.WebMessages
      </base-name>
      <var>bundle</var>
    </resource-bundle>
```

```
<locale-config>
  <default-locale>en</default-locale>
  <!--  <supported-locale>es</supported-locale> -->
</locale-config>
</application>
</faces-config>
```

(There should be no line break and no spaces after `javaee/`.) There is nothing special about this file; just make sure you replace `book.jakartapro.formbasedauth.web` with the folder inside **src/main/resources**, where your language resources are situated. For this example project, you can leave it as shown here.

For the `glassfish-web.xml` file, for now just write the following:

```
<?xml version="1.0" encoding="UTF-8"?>
<glassfish-web-app error-url="">
    <class-loader delegate="true"/>
</glassfish-web-app>
```

The important configuration part gets written into `web.xml`. Here, you define a security constraint and a login configuration, as follows:

```
<?xml version="1.0" encoding="UTF-8"?>
<web-app
  xmlns:xsi=
    "http://www.w3.org/2001/XMLSchema-instance"
  xmlns=
    "http://xmlns.jcp.org/xml/ns/javaee"
  xsi:schemaLocation=
    "http://xmlns.jcp.org/xml/ns/javaee
    http://xmlns.jcp.org/xml/ns/javaee/web-app_4_0.xsd"
  id="WebApp_ID"
  version="4.0">
  <welcome-file-list>
    <welcome-file>index.xhtml</welcome-file>
  </welcome-file-list>
  <servlet>
    <servlet-name>Faces Servlet</servlet-name>
```

```xml
  <servlet-class>
    jakarta.faces.webapp.FacesServlet
  </servlet-class>
  <load-on-startup>1</load-on-startup>
</servlet>
<servlet-mapping>
  <servlet-name>Faces Servlet</servlet-name>
  <url-pattern>*.xhtml</url-pattern>
</servlet-mapping>

<security-constraint>
    <display-name>Admin Constraint</display-name>
    <web-resource-collection>
      <web-resource-name>members</web-resource-name>
      <description />
      <url-pattern>/admin/*</url-pattern>
    </web-resource-collection>
    <auth-constraint>
      <description />
      <role-name>admin</role-name>
    </auth-constraint>
    <user-data-constraint>
      <transport-guarantee>
        CONFIDENTIAL
      </transport-guarantee>
    </user-data-constraint>
</security-constraint>

<login-config>
    <auth-method>FORM</auth-method>
    <realm-name>file</realm-name>
    <form-login-config>
        <form-login-page>/login/login.xhtml</form-login-page>
        <form-error-page>/login/error.xhtml</form-error-page>
    </form-login-config>
</login-config>
```

```
<security-role>
    <description/>
    <role-name>admin</role-name>
</security-role>
</web-app>
```

Because of the settings inside `<security-constraint>`, access to some resources is intercepted by the server, as follows:

- The pattern inside `<url-pattern>` determines which resources to watch. Here, you indicate that all resources underneath `admin/` must be set under security custody.

- The transport protocol is forced to use SSL/TLS. This is because of the `CONFIDENTIAL` note inside `<transport-guarantee>`. A switch from HTTP to HTTPS happens automatically if it's necessary.

- The user has authenticated themself and has the `admin` role. This is because of the `<role-name>` tag.

The `<login-config>` element indicates that you want to use form-based authentication and specifies the login form files. Those files are discussed shortly. Inside `<login-config>`, the `<auth-method>` FORM `</auth-method>` (the FORM being a predefined constant) prescribes a custom form for authentication. The `<realm-name>` file `</realm-name>` indicates the authorization data location. This is server specific—the file here (not a predefined constant as seen in the Jakarta EE specification) is for GlassFish and specifies the server's local file storage for users and passwords.

The `<security-role>` element lists all the roles (apart from unauthorized access) that affect the application. Although you can have several, there is just one in this example.

Security Role Mapping

The server maps the roles (defined in `web.xml`) to the user groups (defined in the server product). This mapping is specified in the WEB-INF / *-web.xml file. For GlassFish, you rewrite the `glassfish-web.xml` file inside the WEB-INF folder and let it read as follows:

```
<?xml version="1.0" encoding="UTF-8"?>
<glassfish-web-app error-url="">
    <security-role-mapping>
        <role-name>admin</role-name>
        <group-name>ApplAdmin</group-name>
    </security-role-mapping>

    <class-loader delegate="true"/>
</glassfish-web-app>
```

The admin role is mapped to the ApplAdmin user group because you used ApplAdmin in the previous sections as the user's group.

Form-Based Authentication XHTML Code

The steps necessary for a web user to enter a secured application part depends on your application, more precisely on the <url-pattern> tag inside the <security-constraint> of the web.xml file. If you, for example, need a button or link to enter the secured area, you write something like this:

```
...
<h:form>
    ...
    <h:button value="Go to secured area"
        outcome="admin/secured.xhtml" />
    ...or...
    <h:outputLink value="admin/secured.xhtml">
        <h:outputText value="Go to secured area"/>
    </h:outputLink>
    ...
</h:form>
```

Note   For brevity, this example uses literal values for the XHTML text. For production code, you should refer to the language resource files instead.

If, in the <url-pattern> tag, inside <security-constraint>, you write the following:

<url-pattern>/admin/*</url-pattern>**

The server knows it has to ask for credentials to enter the secured area given in the button or link.

For the form-based authentication code, login.xhtml is referred to by <login-config> inside the web.xml file. You could write a login form as follows:

```
<!DOCTYPE html>
<html xmlns:h="http://xmlns.jcp.org/jsf/html"
  xmlns:f="http://xmlns.jcp.org/jsf/core"
  xmlns:ui="http://java.sun.com/jsf/facelets"
  xmlns:pt="http://xmlns.jcp.org/jsf/passthrough">
<h:head>
  <title>Login Form</title>
</h:head>
<h:body>
<h1>Please log in:</h1>
<form method="post" action="j_security_check">
  <table columns="2" role="presentation">
    <tr>
      <td><h:outputLabel for="j_username"
                value="User name:"/></td>
      <td><h:inputText id="j_username" autocomplete="off"
                size="30" /></td>
    </tr><tr>
      <td><h:outputLabel for="j_password"
                value="Password:"/></td>
      <td><h:inputSecret id="j_password" autocomplete="off"
                size="30"/></td>
    </tr>
  </table>
  <div class="btnRow">
    <h:commandButton type="submit" styleClass="submitBtn"
          value="Submit"/>
    <h:commandButton type="reset" styleClass="resetBtn"
```

```
        value="Reset"/>
  </div>
</form>
</h:body>
</html>
```

You can code whatever you like, add images and styling as defined by company guidelines, as well as add "Forgot Password" links, information page links, addresses, phone numbers, and more. The important part is this:

```
<form method="post" action="j_security_check">
    <h:inputText id="j_username"/>
    <h:inputSecret id="j_password"/>
</form>
```

Do *not* use <h:form> here. Instead, use the plain HTML <form> tag, as shown here. You have to use j_security_check for the action and j_username and j_password for the field IDs.

For the error page (login/error.xhtml), which will be shown if an invalid user/password combination is entered, you can use the following:

```
<!DOCTYPE html>
<html xmlns:h="http://xmlns.jcp.org/jsf/html"
xmlns:f="http://xmlns.jcp.org/jsf/core"
xmlns:ui="http://java.sun.com/jsf/facelets"
xmlns:pt="http://xmlns.jcp.org/jsf/passthrough">
<head>
<title>Login Error</title>
</head>
<body>
<h1>Invalid user name or password.</h1>

<p>Please enter a username or password that is authorized
    to access the secured part. For this application, this
    means a user that has been created in the
    <code>file</code> realm and has been assigned to the
    <em>group</em> of <code>ApplAdmin</code>.</p>
<p>
```

```
<h:outputLink value="#{request.contextPath}/index.xhtml">
    <h:outputText value="Return to login page"/>
</h:outputLink>
</p>

</body>
</html>
```

Note   This example prepends #{request.contextPath} to the link's value, because the error page cannot use relative URLs. This is because it is handled by the login module, which discards the application's URL context.

The application is functional now. You can deploy it and point your browser to the following URL to test it:

```
https://localhost:8181/FormBasedAuth/
```

## CHAPTER 13

Client Certificates

Using SSL/TLS on the server side to ensure privacy and data integrity is an extremely common task today. With TLS, SSL is deprecated, and you can be sure that your communication is encrypted, and that the identity of the party belonging to the server is also secured. The server presents a certificate to the client to attest its identity. The identity on the client side, however, is checked by comparing a password from a dialog entry with the password from a database.

If this is not enough for your needs, you can require users of your web application to authenticate themselves via a certificate as well. To achieve this, you use *client-side* certificates. This way, you get improved security, because users cannot pin certificates to their monitors as sometimes happens with passwords, and lending a certificate to someone else also is a different story compared to giving away a password. This chapter talks about such these client certificates.

Preparing Scripting

Before you install client certificates, you need to know how to generate them. This book covers professional web applications, so you'll probably have more than a handful of client certificates. Therefore, it would be nice if you could use a script for mass production. I present you such a script, and because this example uses Groovy as the scripting language, you can administer the client certificates from Linux and Windows boxes.

Note   Don't forget to add the Groovy plugin, as described at the beginning of the book.

First, create a new Gradle and Groovy project in Eclipse: Choose File ➤ New ➤ Other... ➤ Gradle Project and enter `ClientCertificates` as a project name. Once the

© Peter Späth 2023
P. Späth, *Pro Jakarta EE 10*, https://doi.org/10.1007/978-1-4842-8214-4_13

project is created, right-mouse-click the project and choose Configure ➤ Convert to Groovy Project to add explicit Groovy scripting capabilities.

Add a new `scripts` folder to the project, and again right-mouse-click the project and choose Properties ➤ Groovy Compiler. Check the Enable Project Specific Settings checkbox. Do the same for Enable Script Folder Support and add the `scripts` folder (enter `scripts/*` in the dialog box).

Note   To adhere more closely to the standards, you can use `src/main/scripts` as the scripts folder. I use Gradle here for dependency resolution only, so using a `scripts` top-level folder, apart from being less Gradle like, is easier to find in the Project Explorer. But if you prefer, go ahead and use the standard location.

In the `build.gradle` file, add dependencies for the certificate administration library called `Bouncycastle`, as follows:

```
dependencies {
  ...
  implementation 'org.bouncycastle:bcprov-jdk15on:1.64'
}
```

In that same file, make sure you are using Java 1.17:

```
plugins {
  ...
}
sourceCompatibility = 1.17
targetCompatibility = 1.17
```

Then right-mouse-click the project and choose Gradle ➤ Refresh Gradle Project.

Generating Client Certificates

To generate one or more client certificates, add the script file called `scripts/generate.groovy`, as follows:

```
import java.security.spec.RSAPrivateKeySpec
import java.security.spec.RSAPublicKeySpec
```

```
import java.security.*
import java.security.cert.*
import java.time.Instant
import java.time.temporal.ChronoUnit

import org.bouncycastle.jce.provider.BouncyCastleProvider
import org.bouncycastle.cert.jcajce.*
import org.bouncycastle.operator.*
import org.bouncycastle.asn1.x500.X500Name
import org.bouncycastle.operator.jcajce.*
import org.bouncycastle.openssl.*
import org.bouncycastle.openssl.jcajce.JcaPEMWriter

KEYSTORE_PASSWORD = "changeit"
RSA_KEYSIZE = 2048
RSA_SIGN_ALGORITHM = 'SHA256WithRSA'
VALIDITY_DAYS = 3650
USERS = [["MarkHamilton","pw173"], ["LindaGreen","pw518"]]

new File("OUTPUT").deleteDir(); new File("OUTPUT").mkdir()
USERS.each { user0 ->
  def user = user0[0]
  def pw = user0[1]

  def certFull = makeCert(user)
  def cert = certFull.cert
  //println(cert.getClass())
  //println(cert)

  FileWriter fileWriter = new FileWriter(
      new File("OUTPUT" + File.separator +
          "clientCert." + user + ".pem"))
  JcaPEMWriter pemWriter = new JcaPEMWriter(fileWriter)
  pemWriter.writeObject(cert)
  pemWriter.flush()

  X509Certificate[] serverChain = new X509Certificate[1]
  serverChain[0] =  cert
```

```
  def keyStore = KeyStore.getInstance("PKCS12")
  keyStore.load(null, pw.toCharArray())
  keyStore.setEntry(user,
      new KeyStore.PrivateKeyEntry(certFull.priv,
                                        serverChain),
      new KeyStore.PasswordProtection(pw.toCharArray())
  )
  def fos = new java.io.FileOutputStream("OUTPUT" +
      File.separator + "clientCert." + user + ".p12")
  keyStore.store(fos, pw.toCharArray())
}

def makeCert(def commonName) {
  def bcProvider = BouncyCastleProvider.PROVIDER_NAME
  Security.addProvider(
    new org.bouncycastle.jce.provider.
        BouncyCastleProvider())

  def keyPairGenerator =
      KeyPairGenerator.getInstance("RSA", bcProvider)
  keyPairGenerator.initialize(RSA_KEYSIZE)
  def keyPair = keyPairGenerator.generateKeyPair()

  def contentSigner =
      new JcaContentSignerBuilder(RSA_SIGN_ALGORITHM).
      build(keyPair.private)

  def dnName = new X500Name("CN=" + commonName)
  def certSerialNumber = BigInteger.valueOf(
      System.currentTimeMillis())
  def startDate = Instant.now()
  def endDate = startDate.plus(VALIDITY_DAYS,
      ChronoUnit.DAYS)
  def certBuilder = new JcaX509v3CertificateBuilder(
      dnName,
      certSerialNumber,
      Date.from(startDate),
```

```
        Date.from(endDate),
        dnName,
        keyPair.public)
  X509Certificate certificate =
        new JcaX509CertificateConverter().
        setProvider(bcProvider).
        getCertificate(certBuilder.build(contentSigner))

  ["priv":keyPair.private,
   "pub":keyPair.public,
   "cert":certificate]
}
```

If necessary, change the all-capital letters at the top part of the file, especially the USERS list for the usernames and passwords. Then run the script by right-mouse-clicking and choosing Run As ➤ Groovy Script. Update the Package Explorer by pressing F5 and you'll see the certificates in the OUTPUT folder. The PKCS12 files are for installing the certificates on the browsers, and the PEM files are the counterparts that will be registered with the server.

Storing the Client Certificate in the Browser

To install the PKCS12 files on the browsers, you must distribute them with an installation manual to your users. Or, as an alternative, you can apply some provisioning software for that task. The installation procedure depends on the browser you use.

If you choose manual installation and use Firefox, open the preferences and go to the Privacy & Security section. In the Certificates section, click View Certificates, and on the Your Certificates tab, click Import. Select the PKCS12 file. You will then be asked for the password. This was specified in the generator script in the users list.

For Internet Explorer on Windows, just double-click the file to start the import procedure. The password you need to enter is the password specified in the generator script in the users list.

Storing the Client Certificate on the Server

To store the client certificates on the server, use the `.pem` files you generated in the previous sections. It is possible to use a shell script to perform this task, but since you already have a Groovy project at hand, you can just as well use a Groovy script to simplify this work and to improve flexibility (filtering, renaming, and so on).

For the GlassFish server, you would back up the existing `cacerts.jks` keystore file, read all the `.pem` files from the build directory `OUTPUT`, pass them into an in-memory representation of the keystore, and eventually overwrite the `cacerts.jks` file with the updated version:

```
import java.security.*
import java.security.cert.*
import java.time.Instant
import java.time.LocalDateTime
import java.time.format.DateTimeFormatter
import java.time.temporal.ChronoUnit

import javax.security.auth.x500.X500Principal

import org.bouncycastle.asn1.x500.X500Name
import org.bouncycastle.cert.jcajce.*
import org.bouncycastle.jce.provider.BouncyCastleProvider
import org.bouncycastle.openssl.PEMParser
import org.bouncycastle.openssl.jcajce.JcaPEMWriter
import org.bouncycastle.operator.jcajce.*

KEYSTORE_PASSWORD = "changeit"
KEYSTORE_ALIAS_PREFIX = ""
IMPORT_FOLDER = "OUTPUT"
GLASSFISH_INST = "/my/local/glassfish/glassfish7"
GLASSFISH_DOMAIN = "domain1"

Security.addProvider(new org.bouncycastle.jce.provider.
    BouncyCastleProvider())

// Backup old cacerts.jks
def FS = File.separator
```

```
def cacerts = GLASSFISH_INST + FS + "glassfish" + FS +
    "domains" + FS +
    GLASSFISH_DOMAIN + FS + "config" + FS + "cacerts.jks"
def bak = cacerts + "." + LocalDateTime.now().format(
    DateTimeFormatter.ISO_LOCAL_DATE_TIME).
    replaceAll(/[^\d]/, "_")
new File(bak).bytes = new File(cacerts).bytes

// Load current keystore
KeyStore ks = KeyStore.getInstance("JKS")
char[] password = KEYSTORE_PASSWORD.toCharArray()
FileInputStream fis  = new FileInputStream(cacerts)
ks.load(fis, password)
fis.close()

// Fetch and register all new keys
new File("OUTPUT").listFiles().
     grep{ f -> f.name.endsWith(".pem") }.each { f ->
    println(f)
    addPemToStore(f, ks)
}

// Overwrite cacerts.jks file
ks.store(new FileOutputStream(new File(cacerts)),
     password)

def addPemToStore(def f, def keyStore) {
    def p = new PEMParser(new FileReader(f))
    def o = p.readObject()
    X509Certificate x509 =
        new JcaX509CertificateConverter().
        setProvider("BC").getCertificate(o)
    println(x509)
    X500Principal princ = x509.subjectX500Principal
    def name = princ.name.replaceAll(/CN\s*=\s*/,"")
    println(name)

    keyStore.deleteEntry(KEYSTORE_ALIAS_PREFIX + name)
```

```
keyStore.setCertificateEntry(
      KEYSTORE_ALIAS_PREFIX + name, x509)

keyStore.aliases().each {
      alias -> println("A: " + alias) }
}
```

Save this file (for example, as `scripts/importToServer.groovy`). As usual, change the all-capital letter variables at the beginning of this file according to your needs.

Note   The Jakarta EE server product determines where the certificates are stored, so for servers other than GlassFish, other procedures apply. Read the server manual to find out how to install certificates.

Client Certificate Web Applications

The `WEB-INF/web.xml` web deployment descriptor used for web applications that want to use client certificates reads as follows:

```
...
<servlet>
  ...
</servlet>
<servlet-mapping>
  ...
</servlet-mapping>

<security-constraint>
    <display-name>Secured Area</display-name>
    <web-resource-collection>
      <web-resource-name>Members</web-resource-name>
      <description />
      <url-pattern>/admin/*</url-pattern>
    </web-resource-collection>
    <auth-constraint>
```

```
    <description />
    <role-name>admin</role-name>
  </auth-constraint>
  <user-data-constraint>
    <transport-guarantee>
      CONFIDENTIAL
    </transport-guarantee>
  </user-data-constraint>
</security-constraint>

<login-config>
  <auth-method>CLIENT-CERT</auth-method>
  <realm-name>certificate</realm-name>
</login-config>
<security-role>
    <description/>
    <role-name>admin</role-name>
</security-role>
...
```

Observe the following characteristics:

- The `<url-pattern>` tag inside `<web-resource-collection>` describes the parts of the web applications that you want to secure.

- The `<role-name>` tag specifies the name of a role that a user needs to acquire in order to gain access to the secured parts. Later, inside the server-specific deployment descriptor (`glassfish-web.xml` for GlassFish), you will see how to describe a mapping from the roles to the user groups.

- The `<transport-guarantee>` CONFIDENTIAL `</transport-guarantee>` code makes sure you will be switching to HTTPS (SSL/TLS) for the secured areas.

- The `<auth-method>CLIENT-CERT</auth-method>` code is the most important part. In this example, you declare that you want to use client certificates for the login into the secured areas. The subsequent `<realm-name>` certificate `</realm-name>` tag tells the server

167

where the login credentials (the certificates) are stored on the server. The latter is a server specific thing—For GlassFish, the `certificate` realm tells GlassFish to use the registered client certificates.

Mapping from roles to user groups happens in the server-specific deployment descriptor. For GlassFish, this is `WEB-INF/glassfish-web.xml`. This file could read as follows:

```xml
<?xml version="1.0" encoding="UTF-8"?>
<glassfish-web-app error-url="">
    <security-role-mapping>
        <role-name>admin</role-name>
        <group-name>authorized</group-name>
    </security-role-mapping>
    <class-loader delegate="true"/>
</glassfish-web-app>
```

This example says to map role `admin` to user group `authorized`. For other servers, you have to consult their manuals to find out how to map roles to user groups.

Additional GlassFish Configuration

In order for GlassFish to use client certificates, you need to enable this feature. Open the web administrator console at `http://localhost:4848` and choose Configurations ➤ Server-config ➤ HTTP Service ➤ Http Listeners ➤ http-listener-2. On the SSL tab, check the Client Authentication checkbox.

Note that there is no relationship between the user groups and the certificates. This is a server product specific setting. For GlassFish, you need to edit this file:

```
GLASSFISH_INST/glassfish/domains/domain1/config/domain.xml
```

Scroll to

```xml
<auth-realm classname="..." name="certificate"/>
```

and change the entry to read

```xml
<auth-realm classname="..." name="certificate">
    <property name="assign-groups"
```

```
            value="authorized">
      </property>
    </auth-realm>
```

You have to restart the server after this change.

Note   There is a bug in GlassFish 7 that prevents you from administering this entry using the web administrator console. Fortunately, you have to edit this file only once for the `certificate` realm, regardless of any client certificate administration activity for that realm.

Client Certificate Example

As an example, create a web application with the following files inside `src/main/webapp/WEB-INF`:

```
beans.xml:
<?xml version="1.0" encoding="UTF-8"?>
<beans xmlns="https://jakarta.ee/xml/ns/jakartaee"
    xmlns:xsi="http://www.w3.org/2001/XMLSchema-instance"
    xsi:schemaLocation="https://jakarta.ee/xml/ns/jakartaee
        https://jakarta.ee/xml/ns/jakartaee/beans_3_0.xsd"
    bean-discovery-mode="all" version="3.0">
</beans>
```

```
faces-config.xml:
<?xml version="1.0" encoding="UTF-8"?>
<faces-config
    xmlns="https://jakarta.ee/xml/ns/jakartaee"
    xmlns:xsi="http://www.w3.org/2001/XMLSchema-instance"
    xsi:schemaLocation="https://jakarta.ee/xml/ns/jakartaee
        https://jakarta.ee/xml/ns/jakartaee/
        web-facesconfig_4_0.xsd"
    version="4.0">
<application>
```

```
<resource-bundle>
  <base-name>
    book.jakartapro.clientcertauth.web.WebMessages
  </base-name>
  <var>bundle</var>
</resource-bundle>
<locale-config>
  <default-locale>en</default-locale>
  <!-- <supported-locale>es</supported-locale> -->
</locale-config>
</application>
</faces-config>
```

glassfish-web.xml:
```
<?xml version="1.0" encoding="UTF-8"?>
<glassfish-web-app error-url="">
    <security-role-mapping>
        <role-name>admin</role-name>
        <group-name>authorized</group-name>
    </security-role-mapping>
    <class-loader delegate="true"/>
</glassfish-web-app>
```

web.xml:
```
<?xml version="1.0" encoding="UTF-8"?>
<web-app
  xmlns:xsi=
    "http://www.w3.org/2001/XMLSchema-instance"
  xmlns=
    "http://xmlns.jcp.org/xml/ns/javaee"
  xsi:schemaLocation=
    "http://xmlns.jcp.org/xml/ns/javaee
     http://xmlns.jcp.org/xml/ns/javaee/web-app_4_0.xsd"
  id="WebApp_ID"
  version="4.0">
  <welcome-file-list>
```

```
  <welcome-file>index.xhtml</welcome-file>
</welcome-file-list>
<servlet>
  <servlet-name>Faces Servlet</servlet-name>
  <servlet-class>
    jakarta.faces.webapp.FacesServlet
  </servlet-class>
  <load-on-startup>1</load-on-startup>
</servlet>
<servlet-mapping>
  <servlet-name>Faces Servlet</servlet-name>
  <url-pattern>*.xhtml</url-pattern>
</servlet-mapping>
<security-constraint>
    <display-name>Admin Constraint</display-name>
    <web-resource-collection>
      <web-resource-name>members</web-resource-name>
      <description />
      <url-pattern>/admin/*</url-pattern>
    </web-resource-collection>
    <auth-constraint>
      <description />
      <role-name>admin</role-name>
    </auth-constraint>
    <user-data-constraint>
      <transport-guarantee>
        CONFIDENTIAL
      </transport-guarantee>
    </user-data-constraint>
</security-constraint>
<login-config>
  <auth-method>CLIENT-CERT</auth-method>
  <realm-name>certificate</realm-name>
</login-config>
<security-role>
```

```
        <description/>
        <role-name>admin</role-name>
    </security-role>
</web-app>
```

Also create the following XHTML files inside src/main/webapp:

index.xhtml:

```
<!DOCTYPE html>
<html xmlns:h="http://xmlns.jcp.org/jsf/html"
xmlns:f="http://xmlns.jcp.org/jsf/core"
xmlns:ui="http://java.sun.com/jsf/facelets"
xmlns:pt="http://xmlns.jcp.org/jsf/passthrough">
<h:head>
<title>Client Cert Authentication</title>
</h:head>

<h:body>
    <div>
        This is the company website
    </div>
    <div>
        <h:form>
        <h:button value="Go to secured area"
            outcome="admin/secured.xhtml" />
        <h:outputLink value="admin/secured.xhtml">
          <f:param name="backref" value="#{view.viewId}"/>
          <h:outputText value="Go to secured area"/>
        </h:outputLink>
        </h:form>
    </div>
</h:body>
</html>
```

admin/secured.xhtml:

```
<!DOCTYPE html>
<html xmlns:h="http://xmlns.jcp.org/jsf/html"
```

```
xmlns:f="http://xmlns.jcp.org/jsf/core"
xmlns:ui="http://java.sun.com/jsf/facelets"
xmlns:pt="http://xmlns.jcp.org/jsf/passthrough">
<h:head>
  <title>Client Cert Authentication</title>
  <h:outputStylesheet library="css" name="style.css" />
</h:head>
<h:body>
    <div>
        Secured area
    </div>
</h:body>
</html>
```

You can see that the client certificate is working by navigating to the secured area. Also, look at the page properties inside your browser. In Firefox, click the security badge next to the URL address in the header, then choose Connection Not Secure →More Information.

Note   Unless you change the HTTPS listener configuration in GlassFish (and if the project name is `ClientCertAuth`), go to `https://localhost:8181/ ClientCertAuth`.

CHAPTER 14

REST Security

With SPAs (single page web applications), it's impossible to reliably enforce security using the security constraints defined in WEB-INF/web.xml. The reason for this is that these security constraints have been especially tailored to HTTP requests and responses of web pages, not JSON snippets. Also, although annotations exist to mark REST classes and methods as subject to security checks, REST implementations are not required to actually perform those checks. The JAX-RS specification for RESTful services just does not handle security.

As a remedy, a handful of additional specifications explicitly handle REST security: JWT (JSON Web Token), JWS (JSON Web Signature), JWE (JSON Web Encryption), JWK (JSON Web Key), and JWA (JSON Web Algorithms). This set of specifications handles security tokens and signing and encryption of JSON data.

Note JSR-375 tried to standardize the Jakarta EE security techniques. However, the implementation seems to be in a questionable state and you cannot find helpful tutorials and examples on the Internet. I therefore decided to use the aforementioned technologies.

This chapter covers securing REST endpoints using the JJWT (Java JSON Web Token) library. It focuses on browser-to-server communication and therefore limits the coverage to JWT and JWS. All the other REST security specifications might improve security, but they are not easily adoptable to browser-to-server communication. If you are thinking about using those as well, consult the specifications and the JJWT documentation at https://github.com/jwtk/jjwt.

© Peter Späth 2023
P. Späth, *Pro Jakarta EE 10*, https://doi.org/10.1007/978-1-4842-8214-4_14

Security Constraints for REST URLs

Say you have a SPA with one HTML page called index.html and two classes of REST data queries, as follows:

```
/data/unsecured/*
/data/secured/*
```

The following sections show you how to develop a method to apply JWT security to the /data/secured/* kind of REST URLs.

About JSON Web Tokens

JSON Web Tokens (JWTs) are data snippets that assert a number of claims in client-server communication. They are especially useful in RESTful applications in which such a claim could be the user's name, role assignments, and more.

The main idea around generating and checking JWTs to make sure the user gets authenticated is as follows:

1. The user initiates the login process, sending credentials (name and password) to the server. This is done by an AJAX call, typically a POST or a PUT.

2. The server tries to authenticate the user.

3. If the authentication fails, a response *without* a JWT is constructed and the server returns from the call.

4. Otherwise, the server generates a new JWT.

5. The JWT is signed with a secret key, stored along with the username and the signing key in some database. The JWT is returned in the response header.

6. The browser client receives the JWT and stores it with the browser session.

7. With each subsequent AJAX request, the client (browser) sends the JWT with the request header. The server decides whether to check the transmitted JWT and reacts accordingly with a success or failure response.

See Figure 14-1.

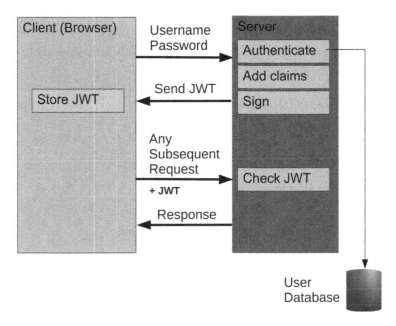

Figure 14-1. *The JWT process*

Note   As is usually the case for security-sensitive applications, the protocol should enforce TLS. In GlassFish, you can disable the unsecured HTTP listener; for other server products, consult the documentation.

Because a secret (private) key is used, the server can be sure the JWT has not been tampered with. From the client (browser) perspective, the JWT is a piece of data that might contain claims like user groups. If the client is not interested in such claims, the JWT is handed around in request and response headers, disregarding any structure or contents.

Preparing GlassFish

Because the coding used in this chapter needs Java runtime permissions that are not automatically granted by GlassFish, you need to make a few adjustments to the server policy. Open the following file:

```
GLASSFISH-INST/glassfish/domains/domain1/
        config/server.policy
```

And add the following lines to it:

```
// Make Jackson serialization work
grant {
  permission java.lang.reflect.ReflectPermission
      "suppressAccessChecks";
};
```

```
// Allow programmatic login
grant {
  permission javax.security.auth.AuthPermission
      "createLoginContext.fileRealm";
  permission javax.security.auth.AuthPermission
      "doAsPrivileged";
};
```

The JWT Login Process, Client Code

For the login process, let's assume you have input fields for the username and password, as well as Login and Logout buttons. In addition, you have two buttons for testing the access to secured and unsecured data, and an area for diagnostic output:

```html
<!DOCTYPE html>
<html>
<head>
<meta charset="UTF-8">
<title>REST Security</title>
<link rel="stylesheet" type="text/css"
  href="css/styles.css" />
```

```html
<script src="js/jquery-3.3.1.min.js"></script>
<script src="js/script.js"></script>
</head>

<body>
  <div>
    User:<input id="user" type="text" />
    Password: <input id="passwd" type="text" />
    <button id="loginButton">Login</button>
    <button id="logoutButton">Logout</button>
  </div>

  <div>
    <button id="accessUnsecuredButton">Access Unsecured</button>
    <button id="accessSecuredButton">Access Secured</button>
  </div>

  <div class="clearfloat"></div>
  <div id="errOutput" class="err"></div>
  <div class="clearfloat"></div>
  <div id="output"></div>
</body>

</html>
```

This is part of the main HTML file. This file and the rest of the client code is inside the webapp source folder:

```
src
  main
    webapp
      static
        index.html
        js
          script.js
          jquery-3.3.1.min.js
        css
          styles.css
```

```
WEB-INF
  web.xml
  glassfish-web.xml
  beans.xml
```

In the `script.js` JavaScript file, you add a variable for the JWT and the code that performs the login and logout activities and reacts to the test buttons:

```javascript
var jwt = '';

$(function() {
  $('#loginButton').click(function(){
    var url = "../jwtlogin/login";
    var loginObj = {
      user: $('#user').val(),
      passwd: $('#passwd').val()
    };
    $.ajax({
      method: "PUT",
      dataType: "json",
      contentType: "application/json",
      url: url,
      data: JSON.stringify(loginObj),
    })
    .done(function(data, status, jqXHR) {
      $('#errOutput').html("");
      $('#output').html("Logged in");
      jwt = jqXHR.getResponseHeader("Authorization");
     })
    .fail(function(jqXHR, textStatus, errorThrown) {
      $('#errOutput').html("AJAX error: " + errorThrown);
      $('#output').html("");
      jwt = "";
    });
  });

  $('#logoutButton').click(function(){
```

```
    var url = "../jwtlogin/logout";
    $.ajax({
      method: "PUT",
      dataType: "json",
      contentType: "application/json",
      url: url
    })
    .done(function(data, status, jqXHR) {
      $('#errOutput').html("");
      $('#output').html("Logged out");
      jwt = jqXHR.getResponseHeader("Authorization");
    })
    .fail(function(jqXHR, textStatus, errorThrown) {
      $('#errOutput').html("AJAX error: " + errorThrown);
      $('#output').html("");
      jwt = "";
    });
    jwt = "";
});

$('#accessUnsecuredButton').click(function(){
  var url = "../unsecured/hello";
  $.ajax({
    method: "GET",
    headers: {Authorization : jwt},
    dataType: "json",
    contentType: "application/json",
    url: url
  })
  .done(function(data, status, jqXHR) {
    $('#output').html(data.msg);
    $('#errOutput').html("");
  })
  .fail(function(jqXHR, textStatus, errorThrown) {
    $('#output').html("");
    $('#errOutput').html("AJAX error: " + errorThrown);
```

```
    });
  });

  $('#accessSecuredButton').click(function(){
    var url = "../secured/hello";
    $.ajax({
      method: "GET",
      headers: {Authorization : jwt},
      dataType: "json",
      contentType: "application/json",
      url: url
    })
    .done(function(data, status, jqXHR) {
      $('#output').html(data.msg);
      $('#errOutput').html("");
    })
    .fail(function(jqXHR, textStatus, errorThrown) {
      $('#output').html("");
      $('#errOutput').html("AJAX error: " + errorThrown);
    });
  });

});
```

The button onclick handlers construct AJAX PUT requests and send them to the server. Result and error messages from the server go to the diagnostic output fields. Upon successful login, the JWT send by the server is stored in the global jwt variable.

The styles.css CSS file for now contains just a few basic styling instructions:

```
body { color: #000044; }
div.clearfloat { clear: both; }
.err { color: red; };
```

The WEB-INF/web.xml file declares the REST servlet. Since you're adding programmatic security, you don't need to add security-related configuration entries here:

```
<?xml version="1.0" encoding="UTF-8"?>
```

```
<web-app
    xmlns:xsi="http://www.w3.org/2001/XMLSchema-instance"
    xmlns="http://xmlns.jcp.org/xml/ns/javaee"
    xsi:schemaLocation="http://xmlns.jcp.org/xml/ns/javaee
        http://xmlns.jcp.org/xml/ns/javaee/web-app_4_0.xsd"
    id="WebApp_ID" version="4.0">
  <display-name>Rest Security</display-name>
  <servlet>
    <servlet-name>jakarta.ws.rs.core.Application
    </servlet-name>
  </servlet>
  <servlet-mapping>
    <servlet-name>jakarta.ws.rs.core.Application
    </servlet-name>
    <url-pattern>/*</url-pattern>
  </servlet-mapping>
</web-app>
```

The beans.xml file specifies all as the discovery mode:

```
<?xml version="1.0" encoding="UTF-8"?>
<beans xmlns="https://jakarta.ee/xml/ns/jakartaee"
    xmlns:xsi="http://www.w3.org/2001/XMLSchema-instance"
    xsi:schemaLocation="https://jakarta.ee/xml/ns/jakartaee
        https://jakarta.ee/xml/ns/jakartaee/beans_3_0.xsd"
    bean-discovery-mode="all" version="3.0">
</beans>
```

The server-specific deployment descriptor glassfish-web.xml can have this basic form:

```
<?xml version="1.0" encoding="UTF-8"?>
<glassfish-web-app error-url="">
    <class-loader delegate="true"/>
</glassfish-web-app>
```

Server Code

The server-side classes go into the book.jakartapro.restsecurity packages:

```
src
  main
    java
      book
        jakartapro
          restsecurity
            StaticContent.java
            NonSecured.java
            Secured.java
            jwt
              JWTLogin.java
              JWTTokenNeeded.java
              JWTTokenNeededFilter.java
              KeyGenerator.java
              JWTKeyGeneratorImpl.java
```

Of course, you can adapt this according to your needs.

The StaticContent class is a simple controller serving the static parts from inside the static/ portion of the web application:

```java
package book.jakartapro.restsecurity;

import java.io.InputStream;

import jakarta.ejb.Stateless;
import jakarta.inject.Inject;
import jakarta.servlet.ServletContext;
import jakarta.ws.rs.GET;
import jakarta.ws.rs.Path;
import jakarta.ws.rs.PathParam;
import jakarta.ws.rs.core.Response;

// http://localhost:8080/RestSecurity/static/index.html
@Path("/static")
```

```
@Stateless
public class StaticContent {

  @Inject ServletContext context;

  @GET
  @Path("/{path : .*}")
  public Response staticResources(
        @PathParam("path") final String path) {
    final InputStream resource = context.
        getResourceAsStream(String.format(
        "/static/%s", path));

    return null == resource ?
        Response.status(Response.Status.NOT_FOUND).build()
      : Response.ok().entity(resource).build();
  }
}
```

The JWT Login Process, Server Code

For the login process, you implement a class called JWTLogin. It is responsible for the jwtlogin/login and jwtlogin/logout REST URLs, and it connects to the container-provided authentication mechanisms:

```
package book.jakartapro.restsecurity.jwt;

import static jakarta.ws.rs.core.HttpHeaders.AUTHORIZATION;
import static jakarta.ws.rs.core.MediaType.
    APPLICATION_JSON;
import static jakarta.ws.rs.core.Response.Status.*;

import java.security.Key;
import java.security.Principal;
import java.time.LocalDateTime;
import java.time.ZoneId;
import java.util.Date;
```

```java
import jakarta.inject.Inject;
import jakarta.servlet.ServletException;
import jakarta.servlet.http.HttpServletRequest;
import jakarta.transaction.Transactional;
import jakarta.ws.rs.Consumes;
import jakarta.ws.rs.PUT;
import jakarta.ws.rs.Path;
import jakarta.ws.rs.Produces;
import jakarta.ws.rs.core.Context;
import jakarta.ws.rs.core.Response;
import jakarta.ws.rs.core.UriInfo;

import io.jsonwebtoken.Jwts;
import io.jsonwebtoken.SignatureAlgorithm;

@Path("/jwtlogin")
@Produces(APPLICATION_JSON)
@Consumes(APPLICATION_JSON)
@Transactional
public class JWTLogin {

  public static class Credentials {
    public String user;
    public String passwd;
    public String getUser() {
      return user;
    }
    public void setUser(String user) {
      this.user = user;
    }
    public String getPasswd() {
      return passwd;
    }
    public void setPasswd(String passwd) {
      this.passwd = passwd;
    }
  }
```

```
@Context
private UriInfo uriInfo;

@Inject
private KeyGenerator keyGenerator;

@PUT
@Path("login")
public Response authenticateUser(Credentials creds,
        @Context HttpServletRequest req) {
  try {
    // Authenticate the user using the credentials
    // provided
    authenticate(creds.user, creds.passwd, req);

    // Issue a token for the user
    String token = issueToken(creds.user);

    // TODO: possibly save token in some database

    // Return the token on the response
    return Response.ok().header(AUTHORIZATION,
        "Bearer " + token).entity("{}").build();

  } catch (Exception e) {
    e.printStackTrace(System.err);
    return Response.status(UNAUTHORIZED).build();
  }
}

@PUT
@Path("logout")
public Response unauthenticateUser(
        @Context HttpServletRequest req) {
  try {
    // TODO: possibly remove token from database

    // Unauthenticate
    unauthenticate(req);
```

```java
      // Return empty response
      return Response.ok().entity("{}").build();
   } catch (Exception e) {
      e.printStackTrace(System.err);
      return Response.status(INTERNAL_SERVER_ERROR).
         build();
   }
}

//////////////////////////////////////////////////////////
//////////////////////////////////////////////////////////

private void authenticate(String login, String password,
      HttpServletRequest req) throws Exception {
   // Authenticate using the container. As an
   // alternative, you could implement
   // your own authentication procedure.
   boolean authenticated = false;
   Principal p = req.getUserPrincipal();
   if(p != null && p.getName().equals(login))
      authenticated = true;
   try {
      if(!authenticated)
         req.login(login, password);
   } catch(ServletException ex) {
      throw new SecurityException("Invalid user/password");
   }
}

private void unauthenticate(HttpServletRequest req)
      throws Exception {
   // Unauthenticate using the container. As an
   // alternative, you could implement
   // your own authentication procedure.
   boolean authenticated = false;
   Principal p = req.getUserPrincipal();
   if(p != null)
```

```
    authenticated = true;
  if(authenticated)
    req.logout();
}

private String issueToken(String login) {
  Key key = keyGenerator.generateKey();
  String jwtToken = Jwts.builder()
      .setSubject(login)
      .setIssuer(uriInfo.getAbsolutePath().toString())
      .setIssuedAt(new Date())
      .setExpiration(toDate(
          LocalDateTime.now().plusMinutes(15L)))
      .signWith(key, SignatureAlgorithm.HS512)
      .compact();
  return jwtToken;
}

private Date toDate(LocalDateTime localDateTime) {
  return Date.from(
    localDateTime.atZone(ZoneId.systemDefault()).
    toInstant());
}
}
```

The characteristics of this class are as follows:

- The class reacts on PUT requests at the https://server.addr:8181/
  RestSecurity/jwtlogin/login and https://server.addr:8181/
  RestSecurity/jwtlogin/logout URLs. Unless otherwise configured,
  RestSecurity corresponds to the name of the WAR file, and 8181 is
  the port of the TLS secured HTTP listener.

- The inner class called Credentials holds credentials. It is needed for
  the REST-enabled authenticateUser() method, such that JAX-RS
  can automatically convert JSON input to a single method parameter.

- The KeyGenerator is used to build a key or sign the JWT. It will never
  be sent to a client.

- In a production environment, the logged-in user, its JWT, and the signing key should all be saved in a database. For brevity, the code just contains TODO marks for that.

- The authenticate() and unauthenticate() methods connect to the container's infrastructure for logging the users in and out.

- The issueToken() method uses the JJWT library to build a JWT.

Sending JWTs Back to the Server

Once a JWT is received in the client-side JavaScript code, it is stored in a local variable. All subsequent AJAX requests are then supposed to transmit this JWT in the Authorization HTTP header.

```
$.ajax({
  method: "GET",
  headers: {Authorization : jwt}, // <===
  dataType: "json",
  contentType: "application/json",
  url: url
})
.done(function(data, status, jqXHR) {
  // Success, access allowed or JWT was ignored...
})
.fail(function(jqXHR, textStatus, errorThrown) {
  // Failure, JWT expired or mismatch...
});
```

You should add the header to *all* AJAX calls, because as for transmitting the JWT, it doesn't matter whether the server will check for the JWT or not. If it doesn't check for it, the header value will just be ignored. If the server checks for it and if the JWT doesn't match or is expired, the server will send an error code with the AJAX response.

Handling JWTs in REST Endpoints

The Secured and NonSecured classes are simple REST endpoint classes that demonstrate secured and unsecured REST access:

```java
package book.jakartapro.restsecurity;

import jakarta.ws.rs.GET;
import jakarta.ws.rs.Path;
import jakarta.ws.rs.PathParam;
import jakarta.ws.rs.Produces;
import jakarta.ws.rs.core.Response;

import book.jakartapro.restsecurity.jwt.JWTTokenNeeded;

@Path("/secured")
@JWTTokenNeeded
public class Secured {
  @GET
  @Path("/{path : .*}")
  @Produces("application/json")
  public Response hello(@PathParam("path") String path) {
    return Response.ok().entity(
        "{\"msg\":\"" + "Secured Content" + "\"}"
        ).build();
  }
}

==--

package book.jakartapro.restsecurity;

import jakarta.ws.rs.GET;
import jakarta.ws.rs.Path;
import jakarta.ws.rs.PathParam;
import jakarta.ws.rs.Produces;
import jakarta.ws.rs.core.Response;
```

```
@Path("/unsecured")
public class NonSecured {
  @GET
  @Path("/{path : .*}")
  @Produces("application/json")
  public Response hello(@PathParam("path") String path) {
    return Response.ok().entity(
        "{\"msg\":\"" + "Unsecured Content" + "\"}"
    ).build();
  }
}
```

The important distinction is that the secured endpoint class is annotated with @ JWTTokenNeeded. This annotation also works at a method level.

The annotation itself uses JAX-WS auto-binding magic. Its code reads as follows:

```
package book.jakartapro.restsecurity.jwt;

import java.lang.annotation.Retention;
import static java.lang.annotation.RetentionPolicy.*;
import java.lang.annotation.Target;
import static java.lang.annotation.ElementType.*;

import jakarta.ws.rs.NameBinding;

@NameBinding
@Retention(RUNTIME)
@Target({TYPE, METHOD})
public @interface JWTTokenNeeded {
}
```

Because of the NameBinding meta-annotation, JAX-WS looks for an implementation of the annotation, which then serves as a filter for REST methods. This implementation does not happen on a Java language basis. Instead, the filter itself is annotated with @ JWTTokenNeeded:

```
package book.jakartapro.restsecurity.jwt;
```

```java
import java.io.IOException;
import java.security.Key;

import jakarta.annotation.Priority;
import jakarta.inject.Inject;
import jakarta.ws.rs.Priorities;
import jakarta.ws.rs.container.ContainerRequestContext;
import jakarta.ws.rs.container.ContainerRequestFilter;
import jakarta.ws.rs.core.HttpHeaders;
import jakarta.ws.rs.core.Response;
import jakarta.ws.rs.ext.Provider;

import io.jsonwebtoken.Jwts;

@Provider
@JWTTokenNeeded
@Priority(Priorities.AUTHENTICATION)
public class JWTTokenNeededFilter
        implements ContainerRequestFilter {

    @Inject
    private KeyGenerator keyGenerator;

    @Override
    public void filter(ContainerRequestContext
            requestContext) throws IOException {
        // Get the HTTP Authorization header from the request
        String authorizationHeader = requestContext.
                getHeaderString(HttpHeaders.AUTHORIZATION);
        if(authorizationHeader == null)
            authorizationHeader = "";

        if(authorizationHeader.equals("")) {
            requestContext.abortWith(Response.status(
                    Response.Status.UNAUTHORIZED).build());
        } else {
            // Extract the token from the HTTP Authorization
            // header
```

```
    String token = authorizationHeader.
        replaceAll("^Bearer ","");
    try {
      // Validate the token
      Key key = keyGenerator.generateKey();
      Jwts.parser().setSigningKey(key).
          parseClaimsJws(token);
    } catch (Exception e) {
      //e.printStackTrace(System.err);
      requestContext.abortWith(Response.status(
          Response.Status.UNAUTHORIZED).build());
    }
  }
 }
}
```

This class extracts the `Authorization` HTTP header and checks whether it contains a valid and non-expired JWT. If everything is fine, it simply returns. Otherwise, the `abortWith(...)` statement causes the AJAX request to return an `UNAUTHORIZED` error code.

In the code, the `keyGenerator.generateKey()` method always returns the same (secret) signing key. In a production environment, you would instead fetch the key from the database. You can get the "user" key for accessing such a database entity, for example, by using the following code:

```
String dd = new String(Base64.getDecoder().
    decode(token.split("\\.")[1]));
JsonReader reader = Json.createReader(
    new StringReader(dd));
JsonObject jsonObject = reader.readObject();
String user = jsonObject.getString("sub");
```

Although this bypasses the JWT signature, it does not impose a security risk if you use this `user` string for no other purpose than to fetch the JWT signature from the database, because this signature could not have been tampered with.

CHAPTER 15

JAVA MVC

Java MVC is a Jakarta EE-compliant instantiation of the model-view-controller pattern. It is an alternative to the component-based Faces technology. The specification of Java MVC is governed by JSR-371, and as an implementation of Java MVC, it uses the Eclipse Krazo project.

About MVC

MVC (model-view-controller) is a rather old architectural pattern originating in 1979. It consists of separating an application into a model layer for the data, a view layer for the presentation on the frontend, and a controller layer for frontend transitions (page flow) and data mediation between the model and the view.

The dedicated web architecture technology for Jakarta EE is Faces, which is not necessarily in contradiction to MVC, but places a bias on a hierarchical component structure, introduces a lifecycle for each request, and lets components dictate the page flow. To go back to its roots and let developers concentrate on MVC and avoid the peculiarities of a component tree, the Java MVC project emerged. The corresponding JSR-371 is in version number 1.0 and it is usable for commercial products. The Eclipse Krazo project (version 2.0.0) provides an implementation.

The specification is available at

```
https://jcp.org/en/jsr/detail?id=371
```

© Peter Späth 2023
P. Späth, *Pro Jakarta EE 10*, https://doi.org/10.1007/978-1-4842-8214-4_15

Installing Java MVC

In order to install Java MVC, you have to add the following dependencies to the Gradle build file:

```
implementation 'jakarta.mvc:jakarta.mvc-api:2.0.0'
implementation 'org.eclipse.krazo:krazo-jersey:3.0.1'
implementation 'jstl:jstl:1.2'
```

The last dependency is not actually needed for MVC, but since you aren't using Faces pages, you have to revert to using JSP pages and add JSTL to allow for improved functionalities like conditional statements and loops in the web page templates. More precisely, the whole build.gradle file reads as follows:

```
plugins {
    id 'war'
}
sourceCompatibility = 1.17
targetCompatibility = 1.17

repositories {
    jcenter()
}

dependencies {
  testImplementation 'junit:junit:4.13.2'

  implementation 'jakarta.platform:jakarta.jakartaee-api:10.0.0'
  implementation 'jakarta.mvc:jakarta.mvc-api:2.0.0'
  implementation 'org.eclipse.krazo:krazo-jersey:3.0.1'
  implementation 'org.glassfish.jersey.core:jersey-server:3.0.0'
  implementation 'jstl:jstl:1.2'
  implementation 'org.webjars:bootstrap:4.3.1'
}

task deployWar(dependsOn: war,
            description:">>> JavaMVC deploy task") {
  doLast {
    def FS = File.separator
```

```
    def glassfish =
        project.properties['glassfish.inst.dir']
    def user = project.properties['glassfish.user']
    def passwd = project.properties['glassfish.passwd']

    File temp = File.createTempFile("asadmin-passwd",
        ".tmp")
    temp << "AS_ADMIN_${user}=${passwd}\n"

    def sout = new StringBuilder()
    def serr = new StringBuilder()
    def libsDir =
      "${project.projectDir}${FS}build${FS}libs"
    def proc = """${glassfish}${FS}bin${FS}asadmin
        --user ${user}
        --passwordfile ${temp.absolutePath}
        deploy --force=true
        ${libsDir}/${project.name}.war""".execute()
    proc.waitForProcessOutput(sout, serr)
    println "out> ${sout}"
    if(serr.toString()) System.err.println(serr)

    temp.delete()
  }
}

task undeployWar(
            description:">>> JavaMVC undeploy task") {
  doLast {
    def FS = File.separator
    def glassfish =
        project.properties['glassfish.inst.dir']
    def user = project.properties['glassfish.user']
    def passwd = project.properties['glassfish.passwd']

    File temp = File.createTempFile("asadmin-passwd",
        ".tmp")
    temp << "AS_ADMIN_${user}=${passwd}\n"
```

```
    def sout = new StringBuilder()
    def serr = new StringBuilder()
    def proc = """${glassfish}${FS}bin${FS}asadmin
        --user ${user}
        --passwordfile ${temp.absolutePath}
        undeploy ${project.name}""".execute()
    proc.waitForProcessOutput(sout, serr)
    println "out> ${sout}"
    if(serr.toString()) System.err.println(serr)

    temp.delete()
  }
}
```

This code adds the usual tasks for local deployment and undeployment. For this to work, add the following to the `gradle.properties` file in the root folder of the Gradle project:

```
glassfish.inst.dir = /path/to/your/glassfish7
glassfish.user = admin
glassfish.passwd =
```

In the rest of this chapter, you learn how to develop a pet store web application as an example project. The project is called `JavaMvc`.

Configuration Files

In `src/main/webapp/WEB-INF`, create two files—a file called `beans.xml`:

```xml
<?xml version="1.0" encoding="UTF-8"?>
<beans xmlns="https://jakarta.ee/xml/ns/jakartaee"
    xmlns:xsi="http://www.w3.org/2001/XMLSchema-instance"
    xsi:schemaLocation="https://jakarta.ee/xml/ns/jakartaee
        https://jakarta.ee/xml/ns/jakartaee/beans_3_0.xsd"
    bean-discovery-mode="all" version="3.0">
</beans>
```

And a file called `glassfish-web.xml`:

```xml
<?xml version="1.0" encoding="UTF-8"?>
<glassfish-web-app error-url="">
    <class-loader delegate="true"/>
</glassfish-web-app>
```

Java MVC is based on JAX-RS and it uses annotations to control the REST API. A web. xml file is therefore not needed. Also, you don't need a traditional index.* (index.jsp or whatever) file for the same reason.

In the main package (called book.jakartapro.javamvc), create an App class with the following contents:

```java
package book.jakartapro.javamvc;

import jakarta.ws.rs.ApplicationPath;
import jakarta.ws.rs.core.Application;

@ApplicationPath("/mvc")
public class App extends Application {
}
```

This adds an additional /mvc to the context root. In the same package, add a web filter named RootRedirector:

```java
package book.jakartapro.javamvc;

import jakarta.servlet.FilterChain;
import jakarta.servlet.annotation.WebFilter;
import jakarta.servlet.http.HttpFilter;
import jakarta.servlet.http.HttpServletRequest;
import jakarta.servlet.http.HttpServletResponse;
import java.io.IOException;

/**
 * Redirecting http://localhost:8080/JavaMvc/
 * This way we don't need a <welcome-file-list> in web.xml
 */
@WebFilter(urlPatterns = "/")
public class RootRedirector extends HttpFilter {
```

```
    private static final long serialVersionUID =
        73329091561636738868L;

    @Override
    protected void doFilter(final HttpServletRequest req,
        final HttpServletResponse res,
        final FilterChain chain) throws IOException {
      res.sendRedirect("mvc/pets");
    }
}
```

This adds a redirector, so that using the base URL in the browser address line directly leads to the pet list.

Starting with Jersey version 2.0, you need to add a bridge between HTML forms and non-HTML REST verbs like DELETE and PUT. In HTML without applying AJAX, issuing DELETE and PUT requests is not possible. On the other hand, since on the server side, you want to stay clean and use a proper application design built on top of JAX-RS, you need a way to simulate DELETE and PUT requests using HTML forms. To this aim, you can install a ContainerRequestFilter and special markers in the HTML forms. The class is named HttpMethodOverride and it reads as follows:

```
package book.jakartapro.javamvc;

import java.io.ByteArrayInputStream;
import java.io.ByteArrayOutputStream;
import java.io.IOException;
import java.io.InputStream;
import java.lang.annotation.Annotation;

import org.glassfish.jersey.message.internal.MediaTypes;

import jakarta.ws.rs.container.ContainerRequestContext;
import jakarta.ws.rs.container.ContainerRequestFilter;
import jakarta.ws.rs.container.PreMatching;
import jakarta.ws.rs.core.Context;
import jakarta.ws.rs.core.Form;
import jakarta.ws.rs.core.MediaType;
import jakarta.ws.rs.ext.Provider;
import jakarta.ws.rs.ext.Providers;
```

```java
@Provider
@PreMatching
public class HttpMethodOverride
      implements ContainerRequestFilter {
  @Context
  private Providers providers;

  @Override
  public void filter(
      ContainerRequestContext requestContext)
      throws IOException {
    if (isForm(requestContext)) {
      Form form = getForm(requestContext);
      String pseudoMethod = (String) form.asMap().
            getFirst("_method");
      if(pseudoMethod != null) {
        requestContext.setMethod(pseudoMethod);
      }
    }

    String methodFromHeader = requestContext.
          getHeaderString("X-HTTP-Method-Override");
    if(methodFromHeader != null) {
      requestContext.setMethod(methodFromHeader);
    }
  }

  //////////////////////////////////////////////////
  //////////////////////////////////////////////////

  private boolean isForm(
      ContainerRequestContext requestContext) {
    return requestContext.hasEntity()
        && MediaTypes.typeEqual(
          MediaType.APPLICATION_FORM_URLENCODED_TYPE,
          requestContext.getMediaType());
  }
```

```java
private Form getForm(
        ContainerRequestContext requestContext)
        throws IOException {
    ByteArrayInputStream bis = toResettableStream(
            requestContext.getEntityStream());
    Form form = providers.getMessageBodyReader(
            Form.class, Form.class, new Annotation[0],
            MediaType.APPLICATION_FORM_URLENCODED_TYPE).
            readFrom(
                Form.class,
                Form.class,
                new Annotation[0],
                MediaType.APPLICATION_FORM_URLENCODED_TYPE,
                null,
                bis);
    bis.reset();
    requestContext.setEntityStream(bis);
    return form;
}

private ByteArrayInputStream toResettableStream(
        InputStream entityStream) throws IOException {
    ByteArrayOutputStream baos = new ByteArrayOutputStream();
    byte[] buffer = new byte[1024];
    int len;
    while ((len = entityStream.read(buffer)) > -1) {
        baos.write(buffer, 0, len);
    }
    baos.flush();
    return new ByteArrayInputStream(baos.toByteArray());
}
}
```

The filter looks for a form parameter called _method, and, if it is present, the filter uses its value to override the REST method. The getForm() and toResettableStream() methods extract the form body and ensure that subsequent filters and the REST method processors get a data stream that hasn't been consumed.

Static Files

You need a very short style file for the Pet Shop application. The application, by virtue of the implementation 'org.webjars:bootstrap:4.3.1' dependency, includes the bootstrap styles. You will see how to use them in the view files. The little addition you need is called styles.css and it goes in the src/main/webapp/staticcss folder. As its contents, write the following:

```
.table td {
  vertical-align: middle;
}
```

For your own development, you can use the same static folder for any static file you need in your application.

Model Classes

The model classes go in a model package. For the sample application, it is called book. jakartapro.javamvc.model and then you add a Pet class:

```
package book.jakartapro.javamvc.model;

import java.util.Objects;
import java.util.UUID;

public final class Pet {
    private UUID petNumber;
    private String name;
    private Status status;

    public Pet() {
        //necessary to fulfill Java Bean Standard
    }

    public Pet(final String name, Status status) {
        this(UUID.randomUUID(), name, status);
    }
```

```java
public Pet(final String name) {
    this(name, Status.ACTIVE);
}

public Pet(final UUID petNumber,
        final String name, final Status status) {
    this.petNumber = petNumber;
    this.name = name;
    this.status = status;
}

public UUID getPetNumber() {
    return petNumber;
}

public void setPetNumber(final UUID petNumber) {
    this.petNumber = petNumber;
}

public String getName() {
    return name;
}

public void setName(final String name) {
    this.name = name;
}

public Status getStatus() {
    return status;
}

public void setStatus(final Status status) {
    this.status = status;
}

public void reserve() {
    this.status = Status.RESERVED;
}
```

```java
    public boolean isReserved() {
        return this.status == Status.RESERVED;
    }

    @Override
    public boolean equals(final Object o) {
        if (this == o) {
            return true;
        }
        if (o == null || getClass() != o.getClass()) {
            return false;
        }
        final Pet pet = (Pet) o;
        return Objects.equals(petNumber, pet.petNumber) &&
                Objects.equals(name, pet.name);
    }

    @Override
    public int hashCode() {
        return Objects.hash(petNumber, name);
    }

    @Override
    public String toString() {
        return "Pet{" +
                "petNumber=" + petNumber +
                ", name='" + name + '\'' +
                '}';
    }

    public enum Status {
        ACTIVE,
        RESERVED
    }
}
```

Controller Classes

The responsibility of the controller class is two-fold: It must define REST entry points that qualify the application as a web application, and it must serve as a mediator between the model and the view layer. For this sample application, the controller class is called `PetshopController`. Move it to a `controller` subpackage, as follows:

```java
package book.jakartapro.javamvc.controller;

import java.util.ArrayList;
import java.util.List;
import java.util.Optional;
import java.util.UUID;

import jakarta.inject.Inject;
import jakarta.mvc.Controller;
import jakarta.mvc.Models;
import jakarta.mvc.binding.BindingResult;
import jakarta.mvc.binding.MvcBinding;
import jakarta.validation.constraints.NotBlank;
import jakarta.validation.constraints.Size;
import jakarta.ws.rs.DELETE;
import jakarta.ws.rs.PUT;
import jakarta.ws.rs.FormParam;
import jakarta.ws.rs.GET;
import jakarta.ws.rs.POST;
import jakarta.ws.rs.Path;
import jakarta.ws.rs.PathParam;
import jakarta.ws.rs.core.Response;
import jakarta.ws.rs.core.Response.Status;

import book.jakartapro.javamvc.model.Pet;
import book.jakartapro.javamvc.service.PetService;

// http://localhost:8080/JavaMvc2/mvc/pets
@Path("/pets")
@Controller
public class PetshopController {
```

```
@Inject
private Models models;

@Inject
private BindingResult bindingResult;

@Inject
private PetService petService;

@GET
public String showIndex() {
    final List<Pet> pets = petService.getAllPets();
    models.put("pets", pets);

    return "index.jsp";
}

@GET
@Path("/{petNumber}")
public Response showDetailsOfPet(
        @PathParam("petNumber") final UUID petNumber) {
    final Optional<Pet> pet = petService.
            findPetForPetNumber(petNumber);

    if (pet.isPresent()) {
        models.put("pet", pet.get());
        return Response.ok("details.jsp")
                .build();
    } else {
        models.put("petNumber", petNumber);
        return Response.status(Status.NOT_FOUND)
                .entity("404.jsp")
                .build();
    }
}

@GET
@Path("/new")
public String showNewPetForm() {
```

```
        return "form.jsp";
    }

    @POST
    public String createNewPet(
            @FormParam("name") @MvcBinding @NotBlank
            @Size(min = 1, max = 255) final String name) {
        if (bindingResult.isFailed()) {
            models.put("errors", new ArrayList<>(
                    bindingResult.getAllErrors()));
            return "form.jsp";
        }

        final Pet pet = petService.createNewPet(name);

        return "redirect:/pets/" + pet.getPetNumber();
    }

    @DELETE
    @Path("/{petNumber}")
    public String deletePet(
            @PathParam("petNumber") final UUID petNumber) {
        petService.deletePet(petNumber);
        return "redirect:/pets";
    }

    @PUT
    @Path("/{petNumber}")
    public String changePet(
            @PathParam("petNumber") final UUID petNumber,
            @FormParam("reserve") @MvcBinding
                    final String reserve) {
        if("reserve".equals(reserve)) {
            petService.reservePet(petNumber);
            return "redirect:/pets";
        }
        return "redirect:/pets";
    }
}
```

The classes from `jakarta.ws.rs` act as you would expect from JAX-RS, with one big difference: A string return will not serve as an outcome to the API call, but instead will map to a view that will determine the HTTP response. The injected `Models` object serves as a container for the model values, and `BindingResult` contains diagnostic information for model values passed from the views to the controller.

As a helper class for a model value container (in a production environment, this is probably a database), add this class:

```
package book.jakartapro.javamvc.repository;

import java.util.ArrayList;
import java.util.Comparator;
import java.util.HashMap;
import java.util.List;
import java.util.Map;
import java.util.Optional;
import java.util.UUID;

import jakarta.annotation.PostConstruct;
import jakarta.ejb.Stateless;

import book.jakartapro.javamvc.model.Pet;

@Stateless
public class PetRepository {
    private final Map<UUID, Pet> petStore =
        new HashMap<>();

    @PostConstruct
    public void init() {
        add(new Pet("Cat"));
        add(new Pet("Dog"));
        add(new Pet("Crocodile", Pet.Status.RESERVED));
        add(new Pet("Snake", Pet.Status.RESERVED));
    }

    public List<Pet> getAll() {
        final List<Pet> pets =
            new ArrayList<>(petStore.values());
```

```
        pets.sort(Comparator.comparing(Pet::getName));

        return pets;
    }

    public Optional<Pet> getByPetNumber(
            final UUID petNumber) {
        return Optional.ofNullable(
                petStore.get(petNumber));
    }

    public void add(final Pet pet) {
        petStore.put(pet.getPetNumber(), pet);
    }

    public boolean update(final Pet pet) {
        return petStore.replace(pet.getPetNumber(), pet)
                != null;
    }

    public void delete(final Pet pet) {
        petStore.remove(pet.getPetNumber());
    }
}
```

The service class mediates between the repository and the controller:

```
package book.jakartapro.javamvc.service;

import java.util.List;
import java.util.Optional;
import java.util.UUID;

import jakarta.ejb.Stateless;
import jakarta.inject.Inject;

import book.jakartapro.javamvc.model.Pet;
import book.jakartapro.javamvc.repository.PetRepository;
```

```java
@Stateless
public class PetService {

    @Inject
    private PetRepository petRepository;

    public List<Pet> getAllPets() {
        return petRepository.getAll();
    }

    public Optional<Pet> findPetForPetNumber(
            final UUID petNumber) {
        return petRepository.getByPetNumber(petNumber);
    }

    public Pet createNewPet(final String name) {
        final Pet pet = new Pet(name);
        petRepository.add(pet);

        return pet;
    }

    public void reservePet(final UUID petNumber) {
        final Optional<Pet> optPet = petRepository.
            getByPetNumber(petNumber);
        if (optPet.isPresent()) {
            final Pet pet = optPet.get();
            pet.reserve();
            petRepository.update(pet);
        }
    }

    public void deletePet(UUID petNumber) {
        final Optional<Pet> optPet = petRepository.
            getByPetNumber(petNumber);
        if (optPet.isPresent()) {
            final Pet pet = optPet.get();
            petRepository.delete(pet);
        }
    }
```

```
    public void setPetRepository(
            final PetRepository petRepository) {
        this.petRepository = petRepository;
    }

}
```

Messages

Just like for any other Jakarta EE-based web application, you can use CDI to inject text messages. This is not Java MVC specific. For the Pet Shop application, use a simple class to access a properties file:

```
package book.jakartapro.javamvc.messages;

import java.util.ResourceBundle;

import jakarta.enterprise.context.RequestScoped;
import jakarta.inject.Inject;
import jakarta.inject.Named;
import jakarta.mvc.MvcContext;

@RequestScoped
@Named("msg")
public class Messages {
  private static final String BASE_NAME = "messages";

  @Inject
  private MvcContext mvcContext;

  public final String get(final String key) {
    final ResourceBundle bundle = ResourceBundle.
        getBundle(BASE_NAME, mvcContext.getLocale());
    return bundle.containsKey(key) ?
        bundle.getString(key) : formatUnknownKey(key);
  }
```

```
  private static String formatUnknownKey(
        final String key) {
    return String.format("???%s???", key);
  }
}
```

The messages reside in src/main/resources/messages.properties:

```
# Petshop application
page.title=MVC Pets

#Header
header.brand=MVC Pets

#Index Page
index.title=Pet overview
index.btn.new=New
index.col.petNumber=Pet number
index.col.name=Name
index.col.status=Status
index.badge.active=Active
index.badge.reserved=Reserved
index.btn.details=Details
index.btn.reserve=Reserve
index.btn.delets=Del

#Form
form.title=New pet
form.label.name=Name
form.label.name.placeholder=Name of pet
form.btn.save=Save

#Details
details.title=Pet details
details.petNumber=Pet number
details.name=Name
details.status=Status

#404
404.title=Couldn't find pet
```

View Pages

The view pages are probably the most challenging part of Java MVC. The reason for this is that:

- You need to know how to communicate with the controller for navigation.

- You need to know how to read and write model data.

- You need to know how to iterate over collections and how to make decisions.

- You need to know how to apply styling.

For Java MVC, you provide JSP pages as view elements inside the `src/main/webapp/WEB-INF/views` folder. This is a convention. It is also possible to use facelets and XHTML template files to that aim. However, since you don't apply a component-based architecture here and some facelet constructs like forms and repetitions lead to problems if they're used inside a Java MVC context, using JSPs and JSTL is a more natural choice.

For the Pet Shop application, you need to provide a `404.jsp` file for any error messages:

```
<%@ page contentType="text/html;charset=UTF-8"
    language="java" %>
<html>
<head>
  [ ... see below ...]
</head>
<body>
<header>
    <nav class="navbar navbar-expand-lg navbar-dark
                bg-dark">
        <div class="container">
<a class="navbar-brand"
   href="${mvc.uri('PetshopController#showIndex')}">
       ${msg.get("header.brand")}
</a>
```

```
        </div>
    </nav>
</header>
<main class="container mt-2">
    <div class="row">
        <h1 class="col-md-12">${msg.get("404.title")}</h1>
        <div class="col-md-12">${petNumber}</div>
    </div>
</main>
</body>
</html>
```

In addition, you need an index.jsp file for the pet list:

```
<%@ page contentType="text/html;charset=UTF-8"
    language="java" %>
<%@ taglib prefix="c"
    uri="http://java.sun.com/jsp/jstl/core" %>
<html>
<head>
    <meta charset="UTF-8">
    <title>${msg.get("page.title")}</title>

    <link rel="stylesheet"
      href="${pageContext.request.contextPath}
      /webjars/bootstrap/4.3.1/css/bootstrap.min.css"/>
    <link rel="stylesheet"
      href="${pageContext.request.contextPath}
      /static/css/styles.css"/>

    <script src="${pageContext.request.contextPath}
      /webjars/jquery/3.0.0/jquery.min.js">
    </script>
    <script src="${pageContext.request.contextPath}
      /webjars/bootstrap/4.3.1/js/bootstrap.bundle.min.js">
    </script>
</head>
```

```html
<body>
<header>
  <nav class="navbar navbar-expand-lg navbar-dark bg-dark">
    <div class="container">
      <a class="navbar-brand"
href="${mvc.uri('PetshopController#showIndex')}">
${msg.get("header.brand")}
      </a>
    </div>
  </nav>
</header>
<main class="container mt-2">
  <div class="row">
    <h1 class="col-md-12">${msg.get("index.title")}</h1>
  </div>
  <div class="row">
    <div class="col-md-12">
      <a class="btn btn-primary"
href="${mvc.uri('PetshopController#showNewPetForm')}">
${msg.get("index.btn.new")}
      </a>
    </div>
  </div>
  <div class="row mt-2">
    <div class="col-md-12">
      <table class="table table-hover">
        <thead>
          <tr>
            <th scope="col">
              ${msg.get("index.col.petNumber")}</th>
            <th scope="col">
              ${msg.get("index.col.name")}</th>
            <th scope="col">
              ${msg.get("index.col.status")}</th>
            <th scope="col"></th>
          </tr>
```

```
      </thead>
      <tbody>
        <c:forEach var="pet" items="${pets}">
          <tr>
            <td>${pet.petNumber}</td>
            <td>${pet.name}</td>
            <td>
              <c:choose>
              <c:when test="${pet.status.name()
                  == 'ACTIVE'}">
                <span class="badge badge-success w-100">
                  ${msg.get("index.badge.active")}
                </span>
              </c:when>
              <c:otherwise>
                <span class="badge badge-danger w-100">
                  ${msg.get("index.badge.reserved")}
                </span>
              </c:otherwise>
              </c:choose>
            </td>
            <td>
              <a class="btn btn-link"
href="${mvc.uriBuilder(
    'PetshopController#showDetailsOfPet').
    build(pet.petNumber)}">${msg.get("index.btn.details")}
              </a>
              <form class="d-inline" method="post"
                    action="${mvc.uriBuilder(
                    'PetshopController#changePet').
                    build(pet.petNumber)}"
                  onsubmit="return confirm(
'Do you really want to reserve the pet?');">
                <input type="hidden" name="_method"
                  value="PUT">
```

```
                    <input type="hidden" name="reserve"
                        value="reserve">
                    <button type="submit"
                        class="btn btn-link">
                      ${msg.get("index.btn.reserve")}
                    </button>
                  </form>
                  <form class="d-inline" method="post"
                        action="${mvc.uriBuilder(
                        'PetshopController#deletePet').
                        build(pet.petNumber)}"
                      onsubmit="return confirm(
'Do you really want to delete the pet?');">
                    <input type="hidden" name="_method"
                        value="DELETE">
                    <button type="submit"
                        class="btn btn-link">
                      ${msg.get("index.btn.delete")}
                    </button>
                  </form>
                </td>
              </tr>
            </c:forEach>
            <c:if test="${pets.isEmpty()}">
              <tr>
                <td>No entries available</td>
              </tr>
            </c:if>
          </tbody>
        </table>
      </div>
    </div>
  </main>
</body>
</html>
```

The details.jsp file reads as follows:

```jsp
<%@ page contentType="text/html;charset=UTF-8"
    language="java" %>
<html>
<head>
  [ ... see below ...]
</head>
<body>
<header>
    <nav class="navbar navbar-expand-lg navbar-dark
                bg-dark">
        <div class="container">
<a class="navbar-brand"
   href="${mvc.uri('PetshopController#showIndex')}">
   ${msg.get("header.brand")}
</a>
        </div>
    </nav>
</header>
<main class="container mt-2">
    <div class="row">
        <h1 class="col-md-12">
          ${msg.get("details.title")}
        </h1>
    </div>
    <hr class="col-md-12">
    <div class="row">
      <div class="col-md-6">
          <label for="petNumber">
            ${msg.get("details.petNumber")}
          </label>
          <input id="petNumber" disabled
              class="form-control"
              value="${pet.petNumber}">
      </div>
```

219

```
        </div>
        <div class="row mt-2">
            <div class="col-md-6">
                <label for="name">
                    ${msg.get("details.name")}
                </label>
                <input id="name" disabled
                    class="form-control"
                    value="${pet.name}">
            </div>
            <div class="col-md-6">
                <label for="status">
                    ${msg.get("details.status")}
                </label>
                <input id="status" disabled
                    class="form-control"
                    value="${pet.status.name()}">
            </div>
        </div>
    </main>
</body>
</html>
```

A last file reads form.jsp as follows:

```
<%@ page contentType="text/html;charset=UTF-8"
    language="java" %>
<%@ taglib prefix="c"
    uri="http://java.sun.com/jsp/jstl/core" %>
<html>
<head>
    [ ... see below ...]
</head>
<body>
<header>
    <nav class="navbar navbar-expand-lg navbar-dark
                bg-dark">
```

```
        <div class="container">
<a class="navbar-brand"
    href="${mvc.uri('PetshopController#showIndex')}">
    ${msg.get("header.brand")}
</a>
        </div>
    </nav>
</header>
<main class="container mt-2">
    <div class="row">
        <h1 class="col-md-12">New pet</h1>
    </div>
    <hr class="col-md-12">
<form method="post"
    action="${mvc.uri('PetshopController#createNewPet')}"
    accept-charset="UTF-8">
        <div class="form-row">
            <div class="col-md-6">
                <label for="name">
                    ${msg.get("form.label.name")}
                </label>
                <input id="name" name="name"
                    class="form-control"
placeholder="${msg.get("form.label.name.placeholder")}"
                    required>
            </div>
        </div>

        <c:choose>
            <c:when test="${not errors.isEmpty()}">
                <div class="row mt-2">
                    <c:forEach var="error"
                                items="${errors}">
                        <div class="col-md-12">
```

221

```
<div class="alert alert-danger">
    ${error.getParamName()}: ${error.getMessage()}
</div>
                            </div>
                        </c:forEach>
                    </div>
                </c:when>
            </c:choose>

            <div class="form-row mt-2">
                <input type="hidden" name="${mvc.csrf.name}"
                        value="${mvc.csrf.token}"/>
                <div class="col-md-3">
                    <button type="submit"
                        class="btn btn-primary">
                            ${msg.get("form.btn.save")}
                    </button>
                </div>
            </div>
        </form>
</main>
</body>
</html>
```

All the view files go to src/main/webapp/WEB-INF/views, and they all have the same <head> part:

```
<head>
    <meta charset="UTF-8">
    <title>${msg.get("page.title")}</title>

    <link rel="stylesheet"
      href="${pageContext.request.contextPath}/webjars/
          bootstrap/4.3.1/css/bootstrap.min.css"/>
    <link rel="stylesheet"
      href="${pageContext.request.contextPath}/static/
          css/styles.css"/>
```

```
<script src="${pageContext.request.contextPath}/
    webjars/jquery/3.0.0/jquery.min.js"></script>
<script src="${pageContext.request.contextPath}/
    webjars/bootstrap/4.3.1/js/
    bootstrap.bundle.min.js"></script>
</head>
```

(Be sure to remove the line breaks and spaces after lines ending with /). A couple of characteristics of the view files are worth mentioning:

- A string like ${petNumber} directly refers to a model value. See for example the end of the 404.jsp file. This model parameter is set in the controller via models.put("petNumber", petNumber);.

- A string like ${msg.get("header.brand")} refers to the Messages class. This is because of its @Named("msg") annotation.

- A string like ${mvc.uri('PetshopController#showIndex')} refers to navigating to a page specified by the result of calling the showIndex() method of the controller. This is described in the JSR-371 specification.

- The /webjars URL part inside <head></head> comes from the bootstrap library. This is not specific to Java MVC and the details are described by the bootstrap documentation.

Running the Pet Shop Application

With all the files in place (see Figure 15-1) and a local GlassFish server running, you can build and deploy the project by invoking the deployWar Gradle task. (From Eclipse, make sure you once right-mouse-clicked the project. Then choose Gradle ➤ Refresh Gradle Project, Taskbar-Refresh-Tasks and Check-Menu ➤ Show All Tasks in the Gradle Tasks View.)

With the application deployed, you can point your browser to http:// localhost:8080/JavaMvc to see the pet list. You can see the details, change the status of a pet to reserved, and create a new pet. See Figure 15-2.

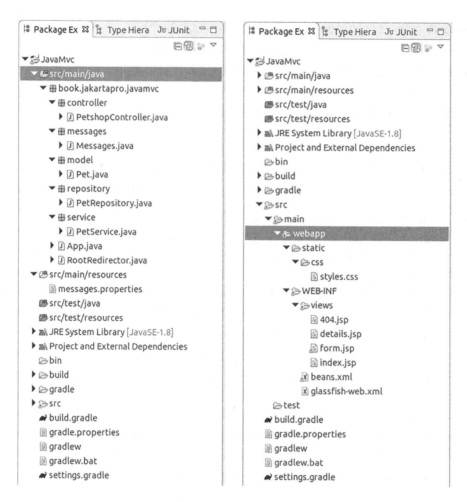

Figure 15-1. *Pet shop sources*

Figure 15-2. *The Pet Shop application*

PART III

Advanced Architecture Related Topics

## CHAPTER 16

Microprofiles

Microservices describe an architectural pattern in which small processor units work together to perform application tasks. Each unit only is responsible for a single service, and all units are loosely coupled and communicate with each other using computer language-independent protocols. In the Java world, the specification set for a microservice is described by the MicroProfile specification, which you can access at the following:

```
https://microprofile.io/
```

Broadly speaking, a microservice yielding the MicroProfile 5.0 specification must be able to handle:

- **JAX-RS 3.0:** For RESTful services

- **Rest Client 3.0:** For REST consumers

- **JSON-P 2.0:** For handling JSON data

- **JSON-B 2.0:** For binding JSON to Java objects

- **CDI 3.0:** For Context and Dependency Injection

- **Jakarta Annotations 2.0:** For annotation processing

- **Config 3.0:** For configuring microservices

- **OpenAPI 3.0:** For REST services metadata describing the services in machine and human readable format

- **Fault Tolerance 4.0:** For improving microservice stability

- **Metrics 4.0:** For letting microservices tell about performance figures

© Peter Späth 2023
P. Späth, *Pro Jakarta EE 10*, https://doi.org/10.1007/978-1-4842-8214-4_16

- **Health 4.0:** For letting microservices tell about their health

- **JWT Authentication 2.0:** For JWT authentication

- **Open Tracing 3.0:** For finding out the relations and communication routes between the microservices

---

Note   The latest MicroProfile version is 6.0. This book uses version 5.0, since this is the version our MicroProfile server adheres to.

At first sight, microservices seem to contradict Jakarta EE, where you have a monolithic architecture with a central server process that hosts dependent modules. The question is, if you are talking about microservices, are you then talking about a competitor to Jakarta EE? Why do I mention microservices in a Jakarta EE book? The answer is surprisingly easy: You can in fact use an application server product with lean memory and lean resources to host a single microservice, thus yielding a microservices architecture based on Jakarta EE! A Jakarta EE microservices landscape is a set of independent, small server processes with a very small number of applications running on each server.

There is one important difference though: The Jakarta EE specification set (EJB, JPA, CDI, etc., versions) and the MicroProfiles specification have different structures. Although you can make a Jakarta EE server follow the MicroProfile specifications, there is no strict conformity of Jakarta EE toward the MicroProfile specification. Nevertheless, thanks to an ongoing effort to coordinate these worlds, MicroProfile 5.0 aligns with Jakarta EE 9, making it easy to switch between those technologies.

If you want to use software that's tailored to follow the MicroProfile specification, a couple of server products follow either approach, including Open Liberty®, Dropwizard®, Thorntail®, Kumuluz®, Payara®, and TomEE®, to name a few.

The rest of this chapter talks about some aspects of TomEE, because it has a commercial-friendly license (Apache license) and is derived from Tomcat, which is a very mature servlet container that has been used in many projects for more than 20 years now. TomEE currently is at version 9.0.0 and adheres to MicroProfile 5.0.

Starting a MicroProfile Sample Project

To begin, you'll use the `RestDate` project from Chapter 2 and enhance it with MicroProfile capabilities. Make a copy of this project and name it `RestDateMP`, or follow the instructions in that chapter and replace `RestDate` with `RestDateMP`. The project structure should read as follows:

```
build.gradle
gradle
|__ wrapper
     |__ gradle-wrapper.jar
     |__ gradle-wrapper.properties
gradlew
gradlew.bat
src
|__ main
     |__ java
     |    |__ book
     |         |__ jakartapro
     |              |__ restdatemp
     |                   |__ RestDate.java
     |__ resources
     |__ webapp
          |__ WEB-INF
               |__ beans.xml
               |__ glassfish-web.xml
               |__ web.xml
```

Start with a standard `build.gradle` file for web applications in Jakarta EE:

```
plugins {
    id 'war'
}
sourceCompatibility = 1.17
targetCompatibility = 1.17
```

```
repositories {
    jcenter()
}

dependencies {
  providedCompile
    'jakarta.platform:jakarta.jakartaee-api:10.0.0'

  // Use JUnit test framework
  testImplementation 'junit:junit:4.13.2'
}
```

For the deployment descriptor called web.xml inside src/main/webapp/WEB-INF, you can write the following:

```xml
<?xml version="1.0" encoding="UTF-8"?>
<web-app xmlns:xsi=
      "http://www.w3.org/2001/XMLSchema-instance"
   xmlns="http://xmlns.jcp.org/xml/ns/javaee"
   xsi:schemaLocation="http://xmlns.jcp.org/xml/ns/javaee
      http://xmlns.jcp.org/xml/ns/javaee/web-app_4_0.xsd"
   id="WebApp_ID" version="4.0">
  <display-name>RestDateMP</display-name>
  <servlet>
    <servlet-name>
      jakarta.ws.rs.core.Application
    </servlet-name>
  </servlet>
  <servlet-mapping>
    <servlet-name>
      jakarta.ws.rs.core.Application
    </servlet-name>
    <url-pattern>/webapi/*</url-pattern>
  </servlet-mapping>
</web-app>
```

Installing a MicroProfile Server

Download the TomEE MicroProfile variant, version 9.0.0 of the TomEE server, from the Downloads section of `http://tomee.apache.org/`. Then unzip the archive to a suitable place.

Changing the Application to a Microservice

To allow for the enhancements that MicroProfile has to offer, you first need to add the MicroProfile dependencies to `build.gradle`:

```
...
dependencies {
  compileOnly 'org.apache.tomee:jakartaee-api:10.0.0'

  providedCompile
    'org.eclipse.microprofile:microprofile:5.0'
  providedCompile
    'org.eclipse.microprofile.openapi:' +
    'microprofile-openapi-api:3.0'

  // Use JUnit test framework
  testImplementation 'junit:junit:4.13.2'

  ...
}
...
```

You can now add MicroProfile capabilities to the application. First, add a few OpenAPI information emitters. A new file called `openapi.yaml` is created and added to `src/main/webapp/META-INF` (be sure to create the folder first):

```
openapi: "3.1.0"
paths:
    /d:
      get:
        description: Get the date
```

```
info:
    title: RestDate
    description: Date operations
    license:
      name: Eclipse Public License - v 1.0
      url: https://www.eclipse.org/legal/epl-v10.html
    version: 1.0.0
servers:
  - url: http://localhost:8080/webapi
```

Adapt the contents according to your needs. In addition, you can add an OpenAPI annotation to the REST class `RestDate.java`:

```java
@Path("/d")
public class RestDate {
  ...
  @GET
  @Produces("application/json")

  // OpenAPI
  @Operation(summary = "Outputs the date",
      description = "This method outputs the date")

  public String stdDate() {
    return "{\"date\":\"" +
        ZonedDateTime.now().toString() + "\"}";
  }
  ...
}
```

OpenAPI can do a lot more. For details, see the OpenAPI specification.

Next, let the application report some performance figures. This is what the Metrics 4.0 specification is about. To use it, add a couple of annotations to the REST method as follows:

```java
  ...
  @GET
  @Produces("application/json")
```

```
// MicroProfile Metrics
  @Timed(name = "stdDate.timer",
      absolute = true,
      displayName = "stdDate Timer",
      description = "Time taken by stdDate.")
  @Counted(name = "stdDate",
      absolute = true,
      displayName = "stdDate call count",
      description = "Number of times we invoked stdDate")
  @Metered(name = "stdDateMeter",
      displayName = "stdDate call frequency",
    description = "Rate the throughput of stdDate.")

// OpenAPI
@Operation(summary = "Outputs the date",
    description = "This method outputs the date")

public String stdDate() {
  return "{\"date\":\"" +
      ZonedDateTime.now().toString() + "\"}";
}
...
```

Invoking this method makes the server process collect the invocation count, elapsed time, and throughput.

The last thing you need to do is add a health check. To this aim, you implement a class that checks whether the format is okay and whether the REST method call leads to an exception. Name this class RestDateHealth and let it read as follows:

```
import org.eclipse.microprofile.health.*;

@Readiness
public class RestDateHealth implements HealthCheck {
  @Override
  public HealthCheckResponse call() {
    HealthCheckResponseBuilder bldr = HealthCheckResponse.
        builder().name("RestDateHealth");
```

```
try {
  String dateStr = new RestDate().stdDate();
  String dayRegex = "\\d{4}-\\d{2}-\\d{2}";
  String tmRegex = "\\d{2}:\\d{2}:\\d{2}\\..*";
  boolean fmtOk = dateStr.matches("\\{\"date\":\"" +
      dayRegex + "T" + tmRegex + "\\}");
  if (fmtOk) {
    return bldr.up().build();
  } else {
    return bldr.down().build();
  }
} catch (Exception e) {
  return bldr.down().withData(
      "Exception", e.getMessage()).build();
}
  }
}
```

The container knows this is a standard health check because of the @Readiness annotation. The contract also implies that the class must implement the HealthCheck interface.

To build the web application WAR archive, invoke the Gradle war task:

```
./gradlew build war
```

```
# Just in case you need a specific Java:
JAVA_HOME=/path/to/the/sdk ./gradlew build war
```

You can also invoke the war task from inside Eclipse in the Gradle Tasks view.

Deploying and Running the Microservice

To deploy the WAR file, copy it from the project to the TomEE server:

```
cp <PROJECT_FOLDER>/build/libs/<PROJECT_NAME>.war \
    <TOMEE_FOLDER>/webapps
```

Substitute suitable values for the uppercase identifiers.

To then start the MicroProfile-enabled TomEE server, enter the following code, again substituting your paths:

```
cd <TOMEE_FOLDER>
bin/startup.sh

# Or, if you want to prescribe a specific Java
cd <TOMEE_FOLDER>
JAVA_HOME=/path/to/the/sdk bin/startup.sh
```

Don't be surprised, as the startup will usually happen extremely quickly—even in just a few seconds.

The logs can be found inside the logs folder, where you can also find the URLs for the various parts of the microservice:

```
# the service itself ->
http://localhost:8080/<APP_NAME>/webapi/d
# see the metrics ->
http://localhost:8080/<APP_NAME>/metrics
# check the health ->
http://localhost:8080/<APP_NAME>/webapi/health
# OpenAPI access ->
http://localhost:8080/<APP_NAME>/openapi
```

The <APP_NAME> usually is the base name of the WAR file (without the suffix).

If you, for example, want to retrieve the health status, enter the following:

```
curl -X GET http://localhost:8080/RestDateMP/webapi/health
```

This implies that you have cURL installed on your system. The output might look like this:

```
{"status":"UP","checks":[
  {"name":"RestDateHealth","status":"UP"},
  {"name":"started-Apache TomEE Server","status":"UP"}
]}
```

In order to get the metrics, enter the following:

```
curl -X GET -H "Accept:application/json" \
  http://localhost:8080/RestDateMP/metrics
```

The output might look like this:

```
{
  "base":{
    "gc.total;name=G1 Young Generation1":11,
    "gc.total;name=G1 Old Generation1":0,
    "cpu.systemLoadAverage":0.16162109375,
    "thread.count":23,
    "classloader.loadedClasses.count":10898,
    "gc.time;name=G1 Old Generation1":0,
    "gc.time;name=G1 Young Generation1":73,
    "classloader.unloadedClasses.total":1,
    "jvm.uptime":517914,
    "thread.max.count":24,
    "memory.committedHeap":201326592,
    "classloader.loadedClasses.total":10899,
    "cpu.availableProcessors":8,
    "thread.daemon.count":20,
    "memory.maxHeap":3116367872,
    "memory.usedHeap":100169072
  },
  "vendor":{
    "memoryPool.CodeHeap 'profiled nmethods'.
        usage":12471168,
    "memoryPool.G1 Old Gen.usage.max":28028416,
    "memoryPool.CodeHeap 'profiled nmethods'.
        usage.max":12473344,
    "memoryPool.CodeHeap 'non-profiled nmethods'.
        usage":3360256,
    "memoryPool.Metaspace.usage.max":56424096,
    "BufferPool_used_memory_mapped":0,
    "memoryPool.CodeHeap 'non-nmethods'.usage.max":1349760,
```

```
    "memoryPool.Compressed Class Space.usage":6693424,
    "memoryPool.G1 Eden Space.usage.max":117440512,
    "memoryPool.G1 Old Gen.usage":28028416,
    "memoryPool.Compressed Class Space.usage.max":6693424,
    "BufferPool_used_memory_mapped -
        'non-volatile memory'":0,
    "memoryPool.G1 Survivor Space.usage":5031792,
    "BufferPool_used_memory_direct":90112,
    "memoryPool.CodeHeap 'non-nmethods'.usage":1292800,
    "memoryPool.Metaspace.usage":56425592,
    "memoryPool.CodeHeap 'non-profiled nmethods'.
        usage.max":3360256,
    "memoryPool.G1 Survivor Space.usage.max":8515800,
    "memoryPool.G1 Eden Space.usage":67108864
},
"application":{
    "stdDate":4,
    "stdDate.timer":{
        "p99":1328098.0,
        "min":127278.0,
        "max":1328098.0,
        "mean":423930.5193750406,
        "p50":143888.0,
        "p999":1328098.0,
        "stddev":502888.22249255574,
        "p95":1328098.0,
        "p98":1328098.0,
        "p75":162595.0,
        "fiveMinRate":0.013113559807197639,
        "fifteenMinRate":0.004419844282431078,
        "meanRate":0.007779616975224711,
        "count":4,
        "oneMinRate":0.06140731004375435,
        "elapsedTime":1761859.0
    },
```

```
"book.jakartapro.restdatemp.RestDate.stdDateMeter":{
  "fiveMinRate":0.013113559807197639,
  "fifteenMinRate":0.004419844282431078,
  "meanRate":0.007779591604845184,
  "count":4,
  "oneMinRate":0.06140731004375435
}
}
```

In order to get the OpenAPI data, enter the following:

```
curl -X GET -H "Accept:application/json" \
    http://localhost:8080/RestDateMP/openapi
```

This produces something like this:

```
{
  "openapi" : "3.1.0",
  "info" : {
    "title" : "RestDate",
    "description" : "Date operations",
    "license" : {
      "name" : "Eclipse Public License - v 1.0",
      "url" : "https://www.eclipse.org/legal/epl-v10.html"
    },
    "version" : "1.0.0"
  },
  "servers" : [ {
    "url" : "http://localhost:8080/webapi"
  } ],
  "paths" : {
    "/d" : {
      "get" : {
        "summary" : "Outputs the date",
        "description" : "This method outputs the date",
        "responses" : {
          "200" : {
```

```
        "description" : "OK",
        "content" : {
          "application/json" : {
            "schema" : {
              "type" : "string"
            }
          }
        }
      }
    }
  }
}
```

For more information, read the specifications at `https://download.eclipse.org/microprofile/`.

CHAPTER 17

Custom CDI

CDI (Context and Dependency Injection) can be used to fetch session or request contexts, inject beans for EL (Expression Language) usage, retrieve entity managers in a JPA context, and more. They do this to seamlessly let classes interact with containers provided by Jakarta EE.

CDI, however, has another facet which regularly gets ignored by introductions, tutorials, and blog entries: You can allow CDI to construct application-internal object graphs or even use your own contexts, producers, interceptors, decorators, stereotypes, and CDI-style events.

CDI Specification

Jakarta EE 10 includes the CDI technology, version 4.0. Find the specification document at

```
https://jakarta.ee/specifications/cdi/4.0
```

This chapter talks about some interesting aspects of CDI for intra-application use. For more information and online resources like the documentation of the reference implementation, Weld can help you. Visit:

```
http://weld.cdi-spec.org/documentation/
```

© Peter Späth 2023
P. Späth, *Pro Jakarta EE 10*, https://doi.org/10.1007/978-1-4842-8214-4_17

Building Object Graphs

With normal Java SE, the *programming model* for building object graphs consists of explicit associations and instantiations of classes:

```java
public class Address {
    private String country;
    private String state;
    private String city;
    private String street;
    private String number;
    private String zip;
    // getters and setters...
}

public class Person {
    private String firstName;
    private String lastName;
    private char gender;
    private String ssn;
    private LocalDate birthday;
    private Address address;
    // getters and setters...
}

public class Invoice {
    private Person buyer;
    private LocalDateTime time;
    private List entries = new ArrayList();
    // getters and setters...
}

...
Person buyer = new Buyer();
buyer.setFirstName("Peter");
// Set other Buyer fields...
Invoice inv = new Invoice();
```

```
inv.setBuyer(buyer);
inv.setTime(LocalDateTime.now());
// Set other Invoice fields...
```

The instantiation and plumbing work shown in the last lines of this snippet go to a suitable method inside your code.

Note   The classes all need to have their own Java files. The rest of this chapter suppresses this note for simplicity.

With CDI, you still declare the associations, but you let the CDI engine instantiate objects and assign them to fields:

```
public class Address {
    private String country;
    private String state;
    private String city;
    private String street;
    private String number;
    private String zip;
    // getters and setters...
}

public class Person {
    private String firstName;
    private String lastName;
    private char gender;
    private String ssn;
    private LocalDate birthday;
    @Inject private Address address;
    // getters and setters...
}

public class Invoice {
    @Inject private Person buyer;
    private LocalDateTime time;
    private List entries = new ArrayList();
```

```
        // getters and setters...
    }

    ...
    // Somewhere, as a class member:
    @Inject Invoice inv;
```

Because of the @Inject, you know that an Invoice object created by CDI *does* have a Person instance in the buyer field (don't use new!), and that the buyer *does* have an Address instance in the address field (again, you don't use new to manually instantiate Address instances).

The most prominent conceptional difference between the non-CDI code and the CDI code is the way an object graph is bootstrapped. With non-CDI, you do all that inside some appropriate method, but with CDI at work, you can just use an injected field @Inject Invoice inv; *inside a class, which itself is under custody of CDI.* This is easy to achieve inside a Faces bean class, a REST endpoint class, or an EJB:

```
    // Faces: -------------------------------------
    @ManagedBean
    @SessionScoped
    public class SomeJsfBean {
      @Inject Invoice inv;

      ...
    }

    // REST endpoint: -----------------------------
    @Path("/invoice")
    public class RestInvoice {
      @Inject Invoice inv;

      @GET
      @Path("show")
      @Produces("application/json")
      public String show() {
        return "{ ... }";
      }
    }
```

```
// EJB: ----------------------------------------
@Stateful
public class InvoiceBean {
    @Inject Invoice inv;
    ...
}
```

By the way, it is also possible to inject instances into constructors and setters:

```
public class SomeClass {
  @Inject
  public SomeClass(Invoice inv) {
      ...
  }
  ...
}
```

```
public class SomeClass {
  ...
  @Inject
  public setInvoice(Invoice inv) {
      ...
  }
  ...
}
```

It is important to understand that in your code you still don't explicitly use those constructors or setters, even with constructor or setter injection at work!

If you programmatically need to fetch an instance of a CDI bean class, you can write the following:

```
import jakarta.enterprise.inject.se.SeContainer;
import jakarta.enterprise.inject.se.SeContainerInitializer;

...
SeContainerInitializer containerInit =
      SeContainerInitializer.newInstance();
SeContainer container = containerInit.initialize();

Invoice proc = container.select(Invoice.class).get();
```

```
...
// Just for debugging / inspection, list all CDI beans
container.getBeanManager().getBeans(Object.class).
    forEach(b -> {
      System.out.println(b);
    });
...

container.close();
```

This is also the way you would use CDI in a non-Jakarta EE environment or in elaborated unit tests.

Qualifiers

CDI can also be used to automatically fetch implementations of interfaces. Consider the following example:

```
public interface Audit {
    ...
}

public class BasicAudit implements Audit {
    ...
}

public class SomeBean {
    @Inject Audit audit;
}
```

CDI sees that BasicAudit is the only implementation of the Audit interface and injects an instance of BasicAudit into the field.

But what happens if you have two implementations, such as in the following setup:

```
public interface Audit {
    ...
}
```

```
public class BasicAudit implements Audit {
    ...
}

public class ComplexAudit implements Audit {
    ...
}

public class SomeBean {
    @Inject Audit audit;
}
```

CDI will now throw an exception, stating that it cannot resolve ambiguous dependencies.

One way to fix this is to mark the different implementations with qualifiers:

```
public interface Audit {
    ...
}

// ------------------------------------------------


import java.lang.annotation.Retention;
import java.lang.annotation.Target;
import static java.lang.annotation.RetentionPolicy.*;
import static java.lang.annotation.ElementType.*;
import jakarta.inject.Qualifier;

@Qualifier
@Retention(RUNTIME)
@Target({TYPE, METHOD, FIELD, PARAMETER})
public @interface Basic {}

// ------------------------------------------------

import java.lang.annotation.Retention;
import java.lang.annotation.Target;
import static java.lang.annotation.RetentionPolicy.*;
import static java.lang.annotation.ElementType.*;
import jakarta.inject.Qualifier;
```

```
@Qualifier
@Retention(RUNTIME)
@Target({TYPE, METHOD, FIELD, PARAMETER})
public @interface Complex {}

// ----------------------------------------------------


@Basic
public class BasicAudit implements Audit {
    ...
}
// ----------------------------------------------------


@Complex
public class ComplexAudit implements Audit {
    ...
}
// ----------------------------------------------------


public class SomeBean {
    @Inject @Complex Audit audit;
    // or: @Inject @Basic Audit audit;
}
```

Such annotations are called *qualifier annotations*. Because of the @Complex (or @Basic) next to the @Inject, CDI now knows which implementation to take.

Note   CDI automatically applies a @Default qualifier to injections if you don't explicitly declare a qualifier. You can suppress this automatic assignment by specifying @Any, which expresses "no qualifier."

Alternatives

To avoid ambiguous dependencies, you can also mark all implementations but one with
@Alternative:

```
public interface Audit {
    ...
}

// --------------------------------------------------

@Default
public class BasicAudit implements Audit {
    ...
}

// --------------------------------------------------

@Alternative
public class ComplexAudit implements Audit {
    ...
}

// --------------------------------------------------

public class SomeBean {
    @Inject Audit audit;
}
```

CDI will select the alternative implementation only if you specify it in WEB-INF/
beans.xml (or META-INF/beans.xml if outside Jakarta EE):

```
<beans
   xmlns="http://xmlns.jcp.org/xml/ns/javaee"
   xmlns:xsi="http://www.w3.org/2001/XMLSchema-instance"
   xsi:schemaLocation="
      http://xmlns.jcp.org/xml/ns/javaee
      http://xmlns.jcp.org/xml/ns/javaee/beans_1_1.xsd"
   bean-discovery-mode="all" version="2.0">
```

```
<alternatives>
  <class>book.jakartapro.cdi.audit.ComplexAudit</class>
</alternatives>
</beans>
```

Or, you can avoid writing something into beans.xml and instead add a @Priority to the alternative implementations:

```
public interface Audit {
    ...
}

// ----------------------------------------------------

@Priority(100) @Alternative
public class BasicAudit implements Audit {
    ...
}

// ----------------------------------------------------

@Priority(101) @Alternative
public class ComplexAudit implements Audit {
    ...
}

// ----------------------------------------------------

public class SomeBean {
    @Inject Audit audit;
}
```

The implementation with the highest @Priority value wins. You can see that, when using the @Priority disambiguation, the example removed the @Default qualifier, since the default does not correspond to a calculable priority.

Producers

If you need more control over the way CDI instantiates classes, or if you need to decide programmatically which implementation of some interface you need for injection purposes, you can provide a *producer method* to create CDI beans:

```java
public interface Audit {
    ...
}

// -------------------------------------------------

@Vetoed
public class BasicAudit implements Audit {
    ...
}

// -------------------------------------------------

@Vetoed
public class ComplexAudit implements Audit {
    ...
}

// -------------------------------------------------

@ApplicationScoped
public class AuditProducer {
  @Produces
  public Audit produce() {
    Audit res = null;
    if(...) { // make a decision
      res = new BasicAudit();
    } else {
      res = new ComplexAudit();
    }
    return res;
  }
}

// -------------------------------------------------
```

```
public class SomeBean {
    @Inject Audit audit;
}
```

This example had to add @Vetoed to the implementations, because otherwise the implementations CDI sees would compete with the producer during the dependency resolution process, yielding an ambiguation exception. Also, the producer class can have any name, but since it is a CDI bean itself and is instantiated by CDI, you need to add @ApplicationScope to the class, which means you at most get one instance of this class for the whole application. Scopes are discussed more in the next section.

Producer methods can have injected parameters, which might help the producer method properly do its work:

```
@Produces
public Audit produce(SomeBean b1,
        SomeOtherBean b2, ...) {
    ...
}
```

Producer methods can have qualifiers assigned to them, just as implementations can have:

```
@ApplicationScoped
public class AuditProducer {
    @Produces @SomeQualifier
    public Audit produce() {
      ...
    }

    @Produces @SomeOtherQualifier
    public Audit produce() {
      ...
    }
}
// -------------------------------------------------
```

```
public class SomeBean {
    @Inject @SomeQualifier Audit audit1;
    @Inject @SomeOtherQualifier Audit audit2;
}
```

The qualifier annotations must be declared exactly the same way as described for implementation qualifiers.

Scope

Scopes determine the lifecycle of beans created by CDI. In a Jakarta EE environment, you learned about a couple of scopes:

- @ApplicationScoped: Any bean in that scope is instantiated only once during the lifetime of an application. All injected beans of the same application refer to the same instance.

- @SessionScoped: Any bean in that scope gets instantiated just once while a user session is active. Any injection occurring in the same user session refers to the same instance.

- @RequestScoped: All request scope beans of the same type refer to the same instance during a request.

- @ConversationScoped: Similar to the session scope, but bound to a user activity inside a session and explicitly controlled by the web application.

If beans don't explicitly declare a scope, they will automatically run inside the @Dependent scope, which means their lifecycle is bound to the object they are injected to. The Dependent scope is called a *pseudo-scope*, since it is derived. There exists a second pseudo-scope named Singleton, which ensures that only one instance of the bean exists on a JVM basis.

You need to add a scope declaration on the type level, or on the method level for producers:

```
@ApplicationScope
public class SomeBean {

    ...

}

// ------------------------------------------------

public class SomeBean {
    @Produces @ApplicationScoped
    public SomeOtherBean produce() {

        ...

    }
}
```

If you are using only intra-app CDI, or you are running outside a Jakarta EE application, only the @ApplicationScope, @Dependent, and Singleton scopes play a role.

If you need to hook into the construction or destruction of a CDI bean, or for debugging purposes, you can use the @PostConstruct and @PreDestroy annotations:

```
public class SomeBean {

    ...

    @PostConstruct
    public void postContruct() {
        System.out.println("POST-CONSTRUCT");
    }

    @PreDestroy
    public void preDestroy() {
        System.out.println("PRE-DESTROY");
    }
}
```

Interceptors

Interceptors are used to add crosscutting concerns to method invocations of CDI beans. This means that you use them to add logging, auditing, performance measurement, and other non-functional code to your bean classes.

CDI interceptors are covered later in this book, in Chapter 18.

Decorators

Decorators are similar to interceptors, in that they somehow extend existing code. But contrary to interceptors, they actually change the functioning of classes, so you use them for things like hotfixes or temporary functional changes.

The only requirement for a CDI bean class to be decoratable is that it must have an interface. As an example, consider the `Processor` interface and the `ProcessorImpl` implementation:

```
public interface Processor {
    void process(Invoice inv);
}

// --------------------------------------------


public class ProcessorImpl implements Processor {
    public void process(Invoice inv) {
        // some code...
    }
}
```

To decorate the processor, all you have to do is write a decorator class as follows:

```
@Decorator
public class ProcessorImplNew implements Processor {
    @Inject @Delegate private ProcessorImpl old;

    public void process(Invoice inv) {
        // some other code...
    }
}
```

Then you declare the decorator inside `WEB-INF/beans.xml` (or `MEA-INF/beans.xml` if you're outside Jakarta EE):

```
<beans...>
  ...

<decorators>
  <class>book.jakartapro.cdi.decorate.ProcessorImplNew</class>
</decorators>
</beans>
```

In the decorator class, you can add a field annotated with `@Inject` `@Delegate`. That means you can access the original implementation, so you could use it for the decorator code.

## CHAPTER 18

Interceptors

Interceptors are methods or classes that hook into method invocations or lifecycle events. You will typically use them for crosscutting concerns like logging, auditing, performance measurements, or any other non-functional aspect of an application. Non-functional means that the program and data flow of the application are not interfered with, so the software functionality stays the same whether or not interceptors are used.

Interceptor targets—classes that are subject to being intercepted—may be session beans, message driven beans, and CDI managed beans. In addition, interceptor targets can also be JPA lifecycle listeners, servlets, and Faces phase listeners. The fact that the latter are not usually called interceptors but more often are called just listeners does not obviate that they can be used the same way as interceptors.

For Jakarta EE 10, the interceptors (not the JPA, servlets, and Faces lifecycle listeners) are described by the Interceptors 2.1 specification.

CDI Interceptors

CDI interceptors are used to add crosscutting concerns to method invocations of CDI beans. You can apply CDI interceptors to *any* class within the custody of CDI. This includes session EJBs, message driven beans, and beans used for Faces and servlets.

To start intercepting, you first declare a new annotation, as in the following snippet:

```
import java.lang.annotation.Retention;
import java.lang.annotation.Target;
import static java.lang.annotation.RetentionPolicy.*;
import static java.lang.annotation.ElementType.*;
import jakarta.interceptor.InterceptorBinding;
```

257

© Peter Späth 2023
P. Späth, *Pro Jakarta EE 10*, https://doi.org/10.1007/978-1-4842-8214-4_18

```
@InterceptorBinding
@Retention(RUNTIME)
@Target({TYPE, METHOD, FIELD})
public @interface TracingInterceptor { }
```

The name is up to you. This example implies that you somehow want to trace the program flow.

The interceptor implementation just needs to add the @Interceptor annotation and the new annotation you just defined:

```
import jakarta.interceptor.*;
import jakarta.annotation.PostConstruct;
import jakarta.annotation.PreDestroy;

@Interceptor
@TracingInterceptor
public class TracingInterceptorSys {
    @AroundInvoke
    public Object logMethodEntry(InvocationContext ctx)
            throws Exception {
        System.out.println("Before entering method: " +
                ctx.getMethod().getName());

        // We call the next interceptor in the
        // interceptor chain.
        return ctx.proceed();
    }

    @AroundConstruct
    public void aroundConstruct(InvocationContext ctx) {
        ...
        ctx.proceed();
    }

    /*or*/
    @AroundConstruct
    public Object aroundConstruct(InvocationContext ctx)
            throws Exception {
```

```
        ...
        return ctx.proceed();
    }

    @AroundTimeout
    public Object aroundTimeout(InvocationContext ctx)
            throws Exception {
        ...
        return ctx.proceed();
    }

    @PostConstruct
    public void postConstruct(InvocationContext ctx) {
        ...
        ctx.proceed();
    }

    @PreDestroy
    public void preDestroy(InvocationContext ctx) {
        ...
        ctx.proceed();
    }
}
```

The annotations can be freely mixed . You are not obliged to use all of them.

In order for the interceptor to be activated, it needs to be added to the WEB-INF/ beans.xml file (or META-INF/beans.xml if it's outside Jakarta EE):

```
<beans...>
  ...

  <interceptors>
    <class>
      book.jakartapro.cdi.intercept.TracingInterceptorSys
    </class>
  </interceptors>
</beans>
```

To apply the interceptor to CDI beans, you have to add the interceptor annotation to a class or method call level:

```
@TracingInterceptor
public class SomeBean {
    ...
}

// --------------------------------------------

public class SomeBean {
    ...
    @TracingInterceptor
    public void someMethod() {
        ...
    }
}
```

Note   This is one of the downsides of CDI interception. The clients (the classes to be intercepted) must explicitly be annotated, which somehow thwarts the non-functional nature of CDI interceptions. Readers of the class code will wonder what this annotation is about and might feel tempted to investigate the non-functional code. Also, it is not possible to add interceptors without touching the code, later in a project, after the development phase. In order to avoid this, you must use other technologies, such as runtime weaving in AOP (aspect oriented programming).

It is possible to avoid creating custom interceptor annotations. Although such custom annotations, like @TracingInterceptor in the previous code, improve the flexibility and the expressiveness and thus readability of the code, you can omit the annotation and directly write the following:

```
import jakarta.interceptor.AroundInvoke;
import jakarta.interceptor.Interceptor;
import jakarta.interceptor.InvocationContext;
```

```
@Interceptor
public class TracingInterceptorSys {
    @AroundInvoke
    public Object logMethodEntry(InvocationContext ctx)
            throws Exception {
        System.out.println("Before entering method: " +
                ctx.getMethod().getName());
        return ctx.proceed();
    }
}

// --------------------------------------------

@Interceptors({TracingInterceptorSys.class})
public class SomeBean {
    ...
}
```

JPA Lifecycle Listeners

You have several options for intercepting JPA lifecycle activities. The first thing you can
do is add methods annotated with @PrePersist, @PostPersist, @PostLoad, @PreUpdate,
@PostUpdate, @PreRemove, and @PostRemove to a JPA entity class (tagged with @Entity).
Such methods can have any name, but they must have a zero-size parameter list and
they must return void.

Second, you can create a special listener class and assign it to a JPA using special
annotations:

```
public class MyListener {
    @PrePersist void onPrePersist(Object o) { ... }
    @PostPersist void onPostPersist(Object o) { ... }
    @PostLoad void onPostLoad(Object o) { ... }
    @PreUpdate void onPreUpdate(Object o) { ... }
    @PostUpdate void onPostUpdate(Object o) { ... }
```

```
  @PreRemove void onPreRemove(Object o) { ... }
  @PostRemove void onPostRemove(Object o) { ... }
}
```

...

```
@Entity @EntityListeners({MyListener.class})
public class TheJpa { ... }
```

The object passed as a parameter is the JPA entity class instance. In addition, you can add such a listener class to *all* JPA entity beans. To this aim, open or create a src/main/resources/META-INF/orm.xml file and add the following:

```xml
<?xml version="1.0" encoding="UTF-8"?>
<entity-mappings
  xmlns="http://java.sun.com/xml/ns/persistence/orm"
  xmlns:xsi="http://www.w3.org/2001/XMLSchema-instance"
  xsi:schemaLocation=
    "http://xmlns.jcp.org/xml/ns/persistence/orm
     http://xmlns.jcp.org/xml/ns/persistence/orm_2_1.xsd"
  version="2.1">
    <persistence-unit-metadata>
        <persistence-unit-defaults>
            <entity-listeners>
                <entity-listener
                    class="the.pac.kage.MyListener" />
            </entity-listeners>
        </persistence-unit-defaults>
    </persistence-unit-metadata>
</entity-mappings>
```

Such a listener frequently gets called a *default entity listener*.

Servlet Listeners

In order to listen to servlet lifecycle events, you create a class as follows:

```
@WebListener()
public class SimpleServletListener implements ... {
    ... implementation ...
}
```

and extend one or more of the following interfaces

```
jakarta.servlet.ServletContextListener
jakarta.servlet.ServletContextAttributeListener
jakarta.servlet.ServletRequestListener
jakarta.servlet.ServletRequestAttributeListener
jakarta.servlet.http.HttpSessionListener
jakarta.servlet.http.HttpSessionAttributeListener
```

These listeners listen to the following events during the lifetime of a servlet:

- `HttpSessionListener`: Listens to HTTP sessions (user sessions) being initialized or destroyed. This is a good hook for auditing purposes.

- `ServletContextListener`: Listens to a servlet's context being initialized or destroyed. You can use this to react to a web application that just started.

- `ServletRequestListener`: Tells about a request being initialized or destroyed.

- `HttpSessionAttributeListener`: Listens to session attributes being added, removed or altered.

- `ServletContextAttributeListener`: Listens to context attributes being added, removed or altered.

- `ServletRequestAttributeListener`: Listens to request attributes being added, removed or altered.

Faces Phase Listeners

To add a phase listener to monitor various Faces phases, you add a listener class declaration to src/main/webapp/WEB-INF/faces-config.xml:

```xml
<?xml version="1.0" encoding="UTF-8"?>
<faces-config
    xmlns="https://jakarta.ee/xml/ns/jakartaee"
    xmlns:xsi="http://www.w3.org/2001/XMLSchema-instance"
    xsi:schemaLocation="https://jakarta.ee/xml/ns/jakartaee
        https://jakarta.ee/xml/ns/jakartaee/
        web-facesconfig_4_0.xsd"
    version="4.0">
    <lifecycle>
      <phase-listener>
          the.pac.kage.MyPhaseListener
      </phase-listener>
      ... more phase listeners ...
    </lifecycle>
    <application>
      ...
    </application>
</faces-config>
```

(Make sure there are no line breaks or spaces after jakartaee/.)

The implementation then indicates which phase it is interested in and presents the methods that are executed before and after the phase:

```java
package the.pac.kage;

...
public class MyPhaseListener implements PhaseListener {
  private static final long serialVersionUID =
      -7607159318721947672L;

  // The phase where the listener is going to be called
  private PhaseId phaseId = PhaseId.RENDER_RESPONSE;
```

```
public void beforePhase(PhaseEvent event) {
  ...
}

public void afterPhase(PhaseEvent event) {
  ...
}

public PhaseId getPhaseId() {
  return phaseId;
}
}
```

Allowed PhaseId values include the following:

1. ANY_PHASE for all phases.

2. RESTORE_VIEW for the *restore view* phase. This is the first phase, where the in-memory view representation is built.

3. APPLY_REQUEST_VALUES for the *apply request values* phase. This is the second phase, where the input field values are read.

4. PROCESS_VALIDATIONS for the validations processing phase.

5. UPDATE_MODEL_VALUES for the fourth phase, where the model is updated.

6. INVOKE_APPLICATION for the fifth phase, where the application is invoked to, for example, find out which page to load next.

7. RENDER_RESPONSE for the last phase, where the response is created.

CHAPTER 19

Bean Validation

If in a JPA entity bean like this

```
@Entity
public class Customer {
  private String name;

  public String getName () {
    return name;
  }

  public void setName (String name) {
    this.name = name;
  }

  ...
}
```

you add a non-null check like this

```
  ...
  public void setName (String name) {
    if(name == null)
      throw new IllegalArgumentException(
        "Name must not be null);
    this.name = name;
  }
  ...
```

You may feel tempted to think of a standard, shorter way to declare such constraints. *Bean validation,* governed by JSR 380, is the answer to this concern. Using bean validation, you can simply write this

© Peter Späth 2023
P. Späth, *Pro Jakarta EE 10*, https://doi.org/10.1007/978-1-4842-8214-4_19

```
...
@NotNull
private String name;

public String getName () {
  return name;
}

public void setName (String name) {
  this.name = name;
}
...
```

The bean validation version inside Jakarta EE 10 is 3.0 and you can see its specification at

```
https://jakarta.ee/specifications/bean-validation/3.0/
    jakarta-bean-validation-spec-3.0.html
```

There are a lot more bean validation constraints. This chapter only talks about some of their usage scenarios.

Where to Use Bean Validation

Bean validation happens at suitable places inside Jakarta EE 10. It is *not* sufficient to just add a constraint annotation to your class for bean validation to work. The framework must also have the class in its custody in order for the constraints to be checked. More precisely, bean validation works at the following places:

- Entity beans (JPA)

- CDI beans

- Faces (beans used in EL constructs, for example during user input fetching)

- JAX-RS (RESTful services)

How to Add Constraints

Constraints can be added to fields as follows:

```
class Person {
  @NotNull @Size(max=30)
  private String name;
}
```

They can be added to constructor and method parameters as follows:

```
class Person {
  public Person(@NotNull @Size(max=30) String name) {
      ...
  }
  ...
  public void setEmail(@Email email) {
      ...
  }
}
```

And they can be added to method return values as follows:

```
class Person {
  ...
  @NotNull @Size(max=30)
  public String getCountry() {
      ...
  }
}
```

Built-in Constraints

All built-in constraints live in the `jakarta.validation.constraints` package. The list is as follows:

- `@NotNull`: Annotated element must not be `null`.

- `@Null`: Annotated element must be `null`.

- @AssertTrue: Annotated element must be a boolean true.

- @AssertFalse: Annotated element must be a boolean false.

- @Min: Annotated element must be a numeric value greater than or equal to some value. The float and double types are not supported, due to possible rounding errors. Example: @Min(37)

- @DecimalMin: Annotated element must be a numeric value greater than or equal to some value specified by a string. The float and double types are not supported, due to possible rounding errors. An optional inclusive parameter specifies whether the specified value is inclusive (the default is true). A null value is considered valid, so you must add NotNull to disallow null. Examples: @DecimalMin("37.4"), @DecimalMin(value="37.4", inclusive=false)

- @Max: Annotated element must be a numeric value less than or equal to some value. The float and double types are not supported, due to possible rounding errors. Example: @Max(42)

- @DecimalMax: Annotated element must be a numeric value less than or equal to some value specified by a string. The float and double types are not supported, due to possible rounding errors. An optional inclusive parameter specifies whether the specified value is inclusive (the default is true). A null value is considered valid, so you must add NotNull to disallow null. Examples: @DecimalMax("42.0"), @DecimalMax(value="42.0", inclusive=false)

- @Negative: Same as @Max(0), but equals is not included and float and double are allowed.

- @Positive: Same as @Min(0), but equals is not included and float and double are allowed.

- @NegativeOrZero: Same as Negative, but 0.0 is included.

- @PositiveOrZero: Same as Positive, but 0.0 is included.

- @Size: The size of a string, collection, map, or array must be between the specified parameters. Optional parameters include min (default is 0) and max (default is Integer.MAX_VALUE). Examples: @Size(min = 5), @Size(min = 10, max = 15)

- @NotEmpty: Same as @Size(min=1).

- @NotBlank: Same as @Size(min=1), only for strings.

- @Digits: Confines the number of integer and fraction digits. Parameters include integer (maximum number of integer digits) and fraction (maximum number of fraction digits).

- @Email: Checks whether a string represents a valid email address. Optional additional parameters include regexp and flags; see @Pattern.

- @Pattern: Annotated element must match the specified pattern. Parameters are regexp (the regular expression) and flags (array of Pattern.Flag values, for regular expression flags). Examples: @Pattern(regexp = "\\d6") (six digits), @Pattern(regexp = "[ABCD]5", flags = Pattern.Flag.CASE_INSENSITIVE) (five from A to D, case-insensitive).

- @Past: The annotated element must represent a date and time in the past. Supported types are Date, Calendar, and more (see the specification).

- @PastOrPresent: Same as @Past, but present is included.

- @Future: The annotated element must represent a date and time in the future. Supported types are Date, Calendar, and more (see the specification).

- @FutureOrPresent: Same as @Future, but present is included.

As a rule of thumb, always add @NotNull if you want to disallow null values, since most constraints consider null a positively validating value.

Custom Constraints

The built-in constraints cover a wide range of validation scenarios. If you still feel tempted to create your own constraint, you have to declare the corresponding annotation and provide an implementation.

As an example, you can create a constraint called PZN8 that denotes the barcode for pharmaceuticals in Germany. It is built from a - sign, plus seven to eight digits, and one check digit at the end. For technical reasons, this string is surrounded by two asterisks (*). You'll want to include the check digit into the validation process, so this is a case in which you have to use a custom validator.

The validation annotation code reads as follows:

```java
package book.jakartapro.restdate;

import static java.lang.annotation.ElementType.*;
import static java.lang.annotation.RetentionPolicy.RUNTIME;

import java.lang.annotation.*;

import jakarta.validation.Constraint;
import jakarta.validation.Payload;

@Target({ METHOD, FIELD, ANNOTATION_TYPE,
          CONSTRUCTOR, PARAMETER, TYPE_USE })
@Retention(RUNTIME)
@Documented
@Constraint(validatedBy = { Pzn8Validator.class })
public @interface PZN8 {
    String message() default "Not a PZN-8";
    Class<?>[] groups() default { };
    Class<? extends Payload>[] payload() default { };

    boolean includeDelimiters() default false;
}
```

For the implementation class called Pzn8Validator, you have to implement the ConstraintValidator interface. In its isValid() method, you take the input string, possibly check and strip off the surrounding asterisks, check for the length, check for and strip off the leading -, check whether all digits are numeric, and then perform the check digit test:

```java
import jakarta.validation.ConstraintValidator;
import jakarta.validation.ConstraintValidatorContext;

public class Pzn8Validator implements
        ConstraintValidator<PZN8, String> {
    protected boolean delimitersIncluded;

    @Override
    public void initialize(PZN8 pzn8) {
        this.delimitersIncluded = pzn8.includeDelimiters();
    }

    @Override
    public boolean isValid(String str,
            ConstraintValidatorContext ctx) {
        // null values are valid
        if ( str == null )
            return true;
        String str2 = str;

        if(delimitersIncluded) {
            if(str.length() < 2) return false;
            if(str.charAt(0) != '*' ||
                str.charAt(str.length()-1) != '*')
                    return false;
            str2 = str.substring(1, str.length() - 1);
        }

        if(str2.length() < 8 || str2.length() > 9)
            return false;

        if(str2.charAt(0) != '-') return false;
        str2 = str2.substring(1);

        char checkDigit = str2.charAt(str2.length() - 1);
        str2 = str2.substring(0, str2.length() - 1);

        if(!str2.matches("\\d*")) return false;
```

```
    // the check digit algorithm for PZNs
    int p = 2;
    int s = 0;
    for(int i=0;i<str2.length();i++) {
        s += p * (str2.charAt(i) - '0');
        p++;
    }
    int x = s % 11;

    return (checkDigit - '0') == x;
  }
}
```

You can now add the custom bean validator to any method, constructor, field, or method parameter. Just as an example inside a REST controller, you could write something like the following:

```
@Path("registerPzn/{pzn}")
@POST
@Produces("application/json")
public Response registerPzn(
    @PathParam("pzn")
    @PZN8 @NotNull String pzn
  )
{
    ... maybe save in database ...
    return Response.status(200)...; // some OK message
}
```

Bean Validation Exceptions

Bean validation issues a jakarta.validation.ValidationException or one of its subclasses when a validation fails. The bean validation specification does not contain any information, because handling this message is the business of the client code.

In RESTful applications using Jax-RS, you would for example add an ExceptionMapper (in the `jakarta.ws.rs.ext` package):

```
import java.util.HashMap;
import java.util.Map;

import jakarta.validation.ConstraintViolation;
import jakarta.validation.ConstraintViolationException;
import jakarta.ws.rs.core.Response;
import jakarta.ws.rs.core.Response.Status;
import jakarta.ws.rs.ext.ExceptionMapper;
import jakarta.ws.rs.ext.Provider;

@Provider
public class ValidationExceptionMapper
        implements
        ExceptionMapper<ConstraintViolationException> {

    @Override
    public Response toResponse(
            ConstraintViolationException ex) {
        Map<String, String> errors =
                new HashMap<String, String>();
        for(ConstraintViolation<?> viol :
                ex.getConstraintViolations()) {
            errors.put(viol.getMessage(),
                    viol.getInvalidValue().toString());
        }
        return Response.status(
                Status.PRECONDITION_FAILED).
            entity(errors).build();
    }
}
```

Note that ConstraintViolationException is a subclass of `jakarta.validation.ValidationException`.

## CHAPTER 20

Jakarta EE Concurrency

Concurrency in Java enterprise applications cannot be achieved the same way as in Java SE, where you can use the classes from the `java.util.concurrent` package and the Thread class and its affiliated classes. The reason for this difference is because the container context data does not automatically propagate to manually created threads, which frequently leads to issues with container-provided resources. Using threads in Jakarta EE's predecessor, JEE, was deprecated until version 8 and developers used to abuse JMS for multithreading, which led to large overhead for concurrency administration and to bad scalability.

Fortunately with JEE version 7 and of course Jakarta EE 8 and later, the situation changed and a dedicated enterprise concurrency API was added. This chapter covers Jakarta EE concurrency using this API.

Preparing the Jakarta EE Server

Concurrency services in Jakarta EE are services provided and managed by the server, so they need to be configured using server administrative tools. For your server product, you have to consult the manual to learn how to do that.

For GlassFish, enter the web administration console at `http://localhost:4848` and choose Resources ➤ Concurrent Resources. You will see the following objects:

- **ContextService**: This service is responsible for context propagation to the threads created by the other services. You normally don't have to alter this service or create more services of this type. The preconfigured default context service is sufficient in most cases. An instance of this service type belongs to the `jakarta.enterprise.concurrent.ContextService` class.

© Peter Späth 2023
P. Späth, *Pro Jakarta EE 10*, https://doi.org/10.1007/978-1-4842-8214-4_20

- **Managed Executor Services:** This service type is most often used when you need concurrency in Jakarta EE. The preconfigured default managed executor service called `defaultManagedExecutorService` can be used as a starting point, and, according to circumstances, your application might be happy with this one. You need its *Logical JNDI Name* (`java:comp/DefaultManagedExecutorService`) in your application. An instance of this service type belongs to the `jakarta.enterprise.concurrent.ManagedExecutorService` class.

- **Managed Scheduled Executor Services:** Same as above, but for scheduled task executions. The *Logical JNDI Name* is `java:comp/DefaultManagedScheduledExecutorService` and the class name is `jakarta.enterprise.concurrent.ManagedScheduledExecutorService`.

- **Managed Thread Factory:** This service type is used for thread creation with propagated container context. This is more low level compared to the other services. You need it if you want to use Java SE concurrency utilities in your application. GlassFish provides one preconfigured managed thread factory with a logical JNDI name of `java:comp/DefaultManagedThreadFactory`.

Using a ManagedExecutorService

A managed executor service will be your entry point if you need high-level concurrency in your Jakarta EE application. There are two ways to acquire this service. The more concise method is to inject it via the following

```
@Resource(name="java:comp/DefaultManagedExecutorService")
private ManagedExecutorService execService;
```

where the resource JNDI name depends on the server product; the name shown corresponds to GlassFish's default and preconfigured service.

The second way to acquire a managed executor service consists of a JNDI lookup:

```
ManagedExecutorService execService =
  InitialContext.
  doLookup("java:comp/DefaultManagedExecutorService");
```

From here, you can register Runnable or Callable tasks to be executed in the background:

```
Future<?> f1 = execService.submit( () -> {
      // not returning anything, corresponds
      // to a Runnable
      ...
   }
);

Future<Integer> f2 = execService.submit( () -> {
      // not returning anything, corresponds
      // to a Runnable
      ...
   }, 42 /* calculated outside the task */
);

Future<Integer> f3 = execService.submit( () -> {
      // returning something, corresponds
      // to a Callable
      ...
      return 42;
   }
);
```

Or even create lists of them:

```
List<Callable<Integer>> retrieverTasks =
     new ArrayList<Callable<Integer>>();
... fill the list ...
List<Future<Integer>> taskResults = execService.
     invokeAll(retrieverTasks);
```

The Future objects you get from the invocation may then be used to wait for the termination of the background task or tasks:

```
...
// waits until background task finishes
f1.get();
...


...
// waits until background task finishes
Integer backgroundRes = f2.get();
...


// waits until all background tasks finish
for (Future<Integer> taskResult : taskResults) {
  try {
    Integer ii = taskResult.get();
    // do something with the res...
  } catch (ExecutionException e) {
    e.printStackTrace(System.err);
  }
}
```

For more details about ManagedExecutorService and for more methods, consult the API documentation. The same holds for Future, which is documented in the Java SE API documentation.

Using a ManagedScheduledExecutorService

A managed scheduled executor service is similar to a managed executor service, except that you can use it for delayed or repeated executions. You acquire the service instance via injection or via JNDI lookup:

```
@Resource(name=
  "java:comp/DefaultManagedScheduledExecutorService")
private ManagedScheduledExecutorService
  schedExecService;
```

...or...

```
ManagedScheduledExecutorService schedExecService =
    InitialContext.
    doLookup(
      "java:comp/DefaultManagedScheduledExecutorService"
    );
```

where the JNDI name is server and it's server-configuration dependent. The name
shown here comes from the preconfigured default service on GlassFish.

To schedule a task for delayed execution, you can now write the following:

```
ScheduledFuture<Integer> sf1 = schedExecService.
    schedule( () -> {
        ...
        return 42;
    }, 5L, TimeUnit.SECONDS);

// or, if you don't need a result:
ScheduledFuture<?> sf2 = schedExecService.
    schedule( () ->
        ...
    }, 5L, TimeUnit.SECONDS);
```

The returned ScheduledFuture object inherits from Future, so you can use it to
cancel the execution or retrieve the result:

```
// In case we want to cancel the execution:
sf1.cancel(true);

...

// Wait for the termination. The sf2 object comes from
// a Runnable, so it can return only 'null' and we don't
// care for the result object.
int res1 = sf1.get();
sf2.get();
```

You can also ask for the remaining timeout value:

```
...
if(!sf1.isDone()) {
  int remainingMillis = sf1.getDelay(
        TimeUnit.MILLISECONDS);
  ...
}
...
```

For repeated executions, you can specify repetition at a fixed rate or specify a fixed delay between the end of one execution and the beginning of the next one:

```
int firstDelay = 5;
int betweenDelay = 2;

ScheduledFuture<?> f1 =
schedExecService.scheduleAtFixedRate(() -> {
      ...
    }, firstDelay, betweenDelay, TimeUnit.SECONDS);

...

ScheduledFuture<?> f2 =
schedExecService.scheduleWithFixedDelay(() -> {
      ...
    }, firstDelay, betweenDelay, TimeUnit.SECONDS);
```

Caution   Be careful with `scheduleAtFixedRate()`. If subsequent task executions take longer than the delay specified, threads may pile up, eventually breaking your server.

A `get()` on the `ScheduledFuture` object returned from the repeated execution scheduling will throw an exception. It doesn't make sense to retrieve the result from a repeated execution. You can, however, use it to cancel the repetition by invoking the `cancel()` method.

Using the ManagedThreadFactory

If you need to go a level deeper and create Thread instances directly inside a Jakarta EE application, you can acquire a jakarta.enterprise.concurrent. ManagedThreadFactory

```
@Resource(name = "java:comp/DefaultManagedThreadFactory")
private ManagedThreadFactory threadFactory;
```

...or...

```
ManagedThreadFactory threadFactory =
    InitialContext.
    doLookup(
      "java:comp/DefaultManagedThreadFactory"
    );
```

The JNDI name is server and it's server-configuration dependent. The name shown here comes from the preconfigured default service on GlassFish.

You can now get a thread by calling the corresponding method of that service:

```
// Thread thr = new Thread(); // NO, NOT IN JAKARTA EE!

Thread thr = threadFactory.newThread( () -> {
        ...
    }
);

thr.start();
```

It is also possible to use ManagedThreadFactory at places where a ThreadFactory is asked for. For example

```
ThreadPoolExecutor exec = new ThreadPoolExecutor(
        5, 10, 5, TimeUnit.SECONDS,
    new ArrayBlockingQueue<Runnable>(10),
    threadFactory);
```

This will give you an ExecutorService that you can then use for concurrency, as described by the Java SE API documentation.

Enterprise Concurrency and Transactions

Transactionality, as defined by JTA, does not propagate from parents to concurrent tasks. It is just too complicated to make sure such an automatism works correctly.

What you can do, however, is to let the background tasks run in a UserTransaction, as follows:

```java
@Resource(name="java:comp/DefaultManagedExecutorService")
private ManagedExecutorService execService;

@Resource
private UserTransaction ut;

...
public void someMethod() {
    Future<?> f1 = execService.submit( new Runnable() {
        public void run() {
            ut.begin();
            try{
                ...
                ut.commit();
            }catch(Exception e){
                ut.rollback();
            }
    }});
    ...
}
```

## CHAPTER 21

Batch Processing

If you need to perform regularly and automatically repeating tasks without user interaction, but based on data that's fed into a database or file by an external process, that's called *batch processing*. For example, consider software that collects employee work hours or aggregates attendance system data each day at midnight and updates employee records in a human resources database. While the first program consists of a user interface in the form of a web application or attendance detection based on cards, badges, or something similar, the aggregation software is a typical candidate for batch processing.

Jakarta EE 10 includes a batch processing technology described by JSR 352. It does *not* include job scheduling or triggering. To this aim, you can use timers, EJBs, or a custom process configured in a `@Singleton @Startup` annotated bean class. Instead, JSR 352 batch processing is a batch programming model, a job specification language, and a batch runtime process. The version of batch processing for Jakarta EE 10 is 2.1.

This chapter covers the Jakarta EE batch processing concepts, without introducing every functionality described by JSR 352. For more details, check out the specification at

```
https://www.jcp.org/en/jsr/detail?id=352
```

Also, the JEE 7 batch processing tutorial at `https://docs.oracle.com/javaee/7/tutorial/batch-processing.htm#GKJIQ6` has valuable information about batch processing, and you can of course find more tutorials on the Internet.

Batch Processing Concepts

Batch processing describes the execution of a number of *batch jobs*. Each job initially runs on a separate thread. A job consists of one or more steps, and by default, the steps run consecutively on the job's single thread. A step may be *partitioned*, which means that its workload will run on separate threads. You can use a partitioning procedure if the items from the input data can be processed independently of each other.

285

© Peter Späth 2023
P. Späth, *Pro Jakarta EE 10*, https://doi.org/10.1007/978-1-4842-8214-4_21

There is a distinction between a *chunk* step type, where the input items get processed, and a *task* step type, which does not see the individual data items but instead is used for initialization, logging, cleanup, or aggregation purposes.

A *flow* is a sequence of steps running as a unit. It is not possible to perform step transitions from individual steps from inside the flow to another step outside the flow. Transitions can only happen after each flow has finished its work. If you want to run several flows in parallel, that's called *split* execution. Flows are optional, so you don't have to use them if you don't need them.

A job may contain *decision* elements, which means the step-by-step workflow may be controlled by decision elements.

Preparing the Server for Batch Processing

To learn whether or not your Jakarta EE server needs to be prepared for batch processing, and how to do this, consult your server manual.

For the GlassFish server, you need to configure a JDBC resource. This is necessary because the server needs a database to store job maintenance data. You can administer this resource for GlassFish (version 7) using the `asadmin` command:

```
./asadmin create-jdbc-connection-pool \
--datasourceclassname \
    org.apache.derby.jdbc.EmbeddedXADataSource \
--restype javax.sql.XADataSource \
--property portNumber=1527:password=APP:user=APP:
    serverName=localhost:
    databaseName=\$\{com.sun.aas.instanceRoot\}/lib/
    databases/batchprocessing:
    connectionAttributes=\;create\\=true \
MyBatchPool
```

(Make sure all the properties are in one line and the spaces are removed.) This installs an embedded Derby database. Derby is included with GlassFish 7, so you don't have to install it. Since it is embedded, you don't have to start it neither. In order to check the connection pool, actually probing the database, enter the following:

```
./asadmin ping-connection-pool MyBatchPool
```

If you ever need to remove it, here's how you do that:

```
./asadmin delete-jdbc-connection-pool MyBatchPool
```

Next you need to configure a JNDI resource to be able to address the new JDBC pool. Enter the following:

```
./asadmin create-jdbc-resource \
    --connectionpoolid MyBatchPool jdbc/MyBatchPool
```

If you need to delete it, write the following:

```
./asadmin delete-jdbc-resource jdbc/MyBatchPool
```

In order to ensure that the batch processing engine uses the new JDBC connection, enter the following in a terminal:

```
./asadmin set-batch-runtime-configuration \
    --datasourcelookupname jdbc/MyBatchPool \
    --executorservicelookupname \
        concurrent/__defaultManagedExecutorService
```

An Employee Attendance Example Batch Processing

As an example for a batch processing, consider a single file collecting employee attendance data as follows:

```
162836945,1459763400000,1459774320000
162836945,1459837380000,1459851300000
618456945,1459756800000,1459783380000
...
```

The first column contains social security numbers, then the start and end times (milliseconds since 1970-01-01 00:00:00 UTC) of an attendance.

You need to write a Jakarta EE enterprise application in form of an EAR file with batch processing acting from inside an EJB as follows:

- In a chunk step, read in the lines from the input file using a *reader*.

- Still in the chunk step, process the items and filter out entries with a missing SSN (social security number).

287

- Still in the chunk step, collect the attendance times and aggregate them for each employee.

- In a subsequent task step, perform any cleanup activities. The input file gets a `.bak` suffix, and empty output folders are possibly deleted.

For simplicity, this example doesn't use a database or write everything into a file system folder, one file per SSN. In a production setup, you would probably enhance this and write everything into a database. Also, the input could as well come from a database instead of a file.

File access in GlassFish is restricted by global permissions. In order for this batch processing application to work, edit the GLASSFISH_INST/glassfish/domains/domain1/config/server.policy file and add this to the end:

```
grant {
    permission java.io.FilePermission
        "<<ALL FILES>>", "write,read,delete";
};
```

Starting a Batch Processing EAR Project

Inside Eclipse, create a new Gradle project. Replace the contents of the build.gradle with the following:

```
plugins {
  id 'ear'
}

// !!!!!!!!!!!!!!!!!!!!!!!!!!!!!!!!!!!!!!!!!!!!!!!!!!!!!!!
// Use the deloyEar task to compile and assemble all and
// upload it to GlassFish
// !!!!!!!!!!!!!!!!!!!!!!!!!!!!!!!!!!!!!!!!!!!!!!!!!!!!!!!

repositories {
    jcenter()
}
```

```
dependencies {
  deploy project(path: ':Batch-Ejb',
      configuration: 'archives')

  // The following dependencies will become EAR libs and
  // will be placed in a dir configured via the libDirName
  // property
  // earlib group: 'log4j', name: 'log4j',
  //     version: '1.2.15', ext: 'jar'
  // earlib 'org.apache.commons:commons-configuration2:2.6'
}

ear {
  //  EAR plugin configuration, if necessary
}

task deployEar(dependsOn: ear,
            description:">>> BATCH deploy task") {
  doLast {
    def FS = File.separator
    def glassfish =
        project.properties['glassfish.inst.dir']
    def user = project.properties['glassfish.user']
    def passwd = project.properties['glassfish.passwd']

    File temp = File.createTempFile("asadmin-passwd",
        ".tmp")
    temp << "AS_ADMIN_${user}=${passwd}\n"

    def sout = new StringBuilder()
    def serr = new StringBuilder()
    def libsDir =
        "${project.projectDir}${FS}build${FS}libs"
    def proc = """"${glassfish}${FS}bin${FS}asadmin
        --user ${user} --passwordfile ${temp.absolutePath}
        deploy --force=true ${libsDir}/${project.name}.ear
        """.execute()
```

```
    proc.waitForProcessOutput(sout, serr)
    println "out> ${sout}"
    if(serr.toString()) System.err.println(serr)

    temp.delete()
  }
}

task undeployEar(
            description:">>> BATCH undeploy task") {
  doLast {
    def FS = File.separator
    def glassfish =
        project.properties['glassfish.inst.dir']
    def user = project.properties['glassfish.user']
    def passwd = project.properties['glassfish.passwd']

    File temp = File.createTempFile("asadmin-passwd",
        ".tmp")
    temp << "AS_ADMIN_${user}=${passwd}\n"

    def sout = new StringBuilder()
    def serr = new StringBuilder()
    def proc = """${glassfish}${FS}bin${FS}asadmin
        --user ${user} --passwordfile ${temp.absolutePath}
        undeploy ${project.name}""".execute()
    proc.waitForProcessOutput(sout, serr)
    println "out> ${sout}"
    if(serr.toString()) System.err.println(serr)

    temp.delete()
  }
}
```

Inside the settings.gradle file, add one line for the EJB subproject:

```
include 'Batch-Ejb'
```

To connect to the local GlassFish server, add the `gradle.properties` file in the project root and let it read as follows:

```
glassfish.inst.dir = /glassfish/inst/dir
glassfish.user = admin
glassfish.passwd =
```

where the `user` and the empty password correspond to a vanilla GlassFish installation. Of course, you must provide your own local GlassFish installation folder as a value for the first property.

For the subproject, create a folder called `Batch-Ejb`, situated as a child to the main project folder. Copy `gradlew` and `gradlew.bat` and the complete `gradle` folder into `Batch-Ejb`. This allows you to build the EJB JAR from inside a terminal using the Gradle wrapper script. Also inside `Batch-Ejb`, create a file called `build.gradle` and let it read as follows:

```
plugins {
    id 'java'
}
sourceCompatibility = 1.17
targetCompatibility = 1.17

repositories {
    jcenter()
}

dependencies {
    testImplementation 'junit:junit:4.13.2'
    compileOnly 'jakarta.platform:jakarta.jakartaee-api:10.0.0'
}
```

You can now right-mouse-click (the root project) and then choose Gradle → Refresh Gradle Project. Eclipse will henceforth show two projects in the Package Explorer view, as shown in Figure 21-1. Finally, add the following folders inside the `Batch-Ejb` subproject:

```
src/main/java
src/main/resources/META-INF
```

Make sure Eclipse recognizes the first one as a source folder (use project properties for that).

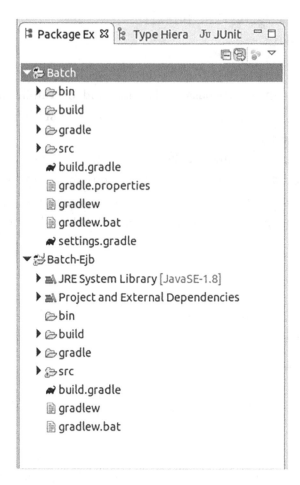

Figure 21-1. *Batch project and subproject*

Note   Despite being shown as two separate projects in Eclipse, the separation into main project and subproject (in a subfolder) still holds.

Example Data Preparation

You can use any suitable folder on your file system as the example data for the employee attendance batch processing application. The following sections call this folder DATA_FOLDER.

Create an input file called DATA_FOLDER/attendance.csv and let it read as follows:

```
162836945,1459763400000,1459774320000
162836945,1459837380000,1459851300000
618456945,1459756800000,1459783380000
```

The columns are Social Security Number, Work Begin Time, and Work End Time (in milliseconds since 1970-01-01 00:00:00 UTC).

Job Definition File

The job definition file is the main configuration file of the batch processing application. It is called myJobName.xml and it's in this folder of the Batch-Ejb subproject:

```
src/main/resources/META-INF/batch-jobs
```

Instead of myJobName, you can use a different name for the job definition file, but you cannot change the .xml suffix.

Create the file and let its contents read as follows:

```xml
<job id="attendanceJob"
    xmlns="https://jakarta.ee/xml/ns/jakartaee"
    version="2.0">

<!-- Properties. Can be read from inside the code    -->
<properties>
    <property name="input_file"
        value="/path/to/DATA_FOLDER/attendance.csv"/>
    <property name="output_folder"
        value="/path/to/DATA_FOLDER/employee"/>
</properties>
```

```
<!-- One chunk step. ............................... -->
<!-- The "next" attribute tells which step to call   -->
<!-- next once this step finishes                     -->
<step id="attendance" next="cleanup">
  <!-- We make this step a junk type step.           -->
  <!-- Because of checkpoint-policy="item" the        -->
  <!-- processing gets committed after each           -->
  <!-- "item-count" processed items.                  -->
  <!-- We define a reader, a processor and a writer   -->
  <chunk checkpoint-policy="item" item-count="10">
    <reader ref="AttendanceReader"></reader>
    <processor ref="AttendanceProcessor"></processor>
    <writer ref="AttendanceWriter"></writer>
  </chunk>
</step>

<!-- One cleanup step. Implemented by the class       -->
<!-- @Named "Cleanup"                                 -->
<!-- Because of the "end on" this is the last step    -->
<!-- of the job.                                      -->
<step id="cleanup">
  <batchlet ref="Cleanup"></batchlet>
  <end on="COMPLETED"/>
</step>
</job>
```

You can see that this example defined two properties for the input file and the output files. You can later use these properties from inside the coding. One chunk type step is for reading, processing, and writing the processed data, and the other task step is for cleaning up.

Checkpoints define points in the processing that can be restarted after a planned or unplanned processing interruption. You can allow at maximum of ten chunk items to be re-read if something happens.

The various ref attributes refer to the CDI @Named classes.

Batch Processing Scheduling

You can use a timer EJB to schedule batch processing. Create any class inside the Batch-Ejb project and let it read as follows:

```java
package book.jakartapro.batch;

import java.util.Properties;

import jakarta.batch.operations.JobOperator;
import jakarta.batch.runtime.BatchRuntime;
import jakarta.ejb.Schedule;
import jakarta.ejb.Stateless;

@Stateless
public class BatchEjb {
  @Schedule(
      // for development and testing, we start the
      // job each 10 seconds
      hour = "*",
      minute = "*",
      second = "*/10",
      persistent = false)
  public void timeout() {
    try {
      JobOperator jobOperator =
          BatchRuntime.getJobOperator();

      // Just in case we want to programmatically add
      // a property
      Properties props = new Properties();
      props.setProperty("parameter1", "value1");

      // The "job1" corresponds to the name of
      // the job definition file (without the .xml)
      // inside META-INF/batch-jobs
      long execID = jobOperator.start("job1", props);
```

```
  } catch (Exception e) {
    e.printStackTrace(System.err);
    ...
  }
 }
}
```

The class name and position (package) is up to you. Jakarta knows this is a timer EJB just by the @Schedule annotation.

Batch Processing Java Artifacts

The various ref attributes inside the job definition file correspond to the following classes. First, define a class that holds an input item:

```
package book.jakartapro.batch;

import java.io.Serializable;

public class AttendanceItem implements Serializable {
  private static final long serialVersionUID =
      2413741734728319063L;
  private String ssn; // + getters/setters
  private int workMillis; // + getters/setters

  public AttendanceItem(String ssn, int workMillis) {
    this.ssn = ssn;
    this.workMillis = workMillis;
  }
}
```

Next, implement the reader in the chunk step:

```
package book.jakartapro.batch;

import java.io.*;
import java.util.Properties;

import jakarta.batch.api.chunk.ItemReader;
import jakarta.batch.runtime.context.JobContext;
```

```java
import jakarta.enterprise.context.Dependent;
import jakarta.inject.Inject;
import jakarta.inject.Named;

@Dependent
@Named("AttendanceReader")
public class AttendanceReader implements ItemReader {
  private Integer lineNumber = 0;
  private String fileName;
  private InputStream ins;
  private BufferedReader breader;

  @Inject
  private JobContext jobCtx;

  public AttendanceReader() {
  }

  @Override
  public void open(Serializable checkpoint)
        throws Exception {
    Properties props = jobCtx.getProperties();
    fileName = props.getProperty("input_file");
    System.out.println("open file " + fileName);

    if (new File(fileName).exists()) {
      ins = new FileInputStream(fileName);
      breader = new BufferedReader(
          new InputStreamReader(ins));

      if (checkpoint != null) {
        lineNumber = (Integer) checkpoint;
        for (int i = 0; i < lineNumber; i++) {
          breader.readLine();
        }
      }
    }
  }
```

```java
  @Override
  public void close() throws Exception {
    if(breader != null) breader.close();
    if(ins != null) ins.close();
  }

  @Override
  public Object readItem() throws Exception {
    Object result = null;
    if(breader == null) return result;

    String line = breader.readLine();
    if (line != null) {
      String[] fields = line.split("[, \t\r\n]+");
      if (fields.length > 0) {
        String ssn = fields[0].trim();
        long tm = Long.parseLong(fields[2]) -
                Long.parseLong(fields[1]);
        AttendanceItem itemObj =
                new AttendanceItem(ssn, (int) tm);
        result = itemObj;
      }
      lineNumber++;
    }
    return result;
  }

  @Override
  public Serializable checkpointInfo() throws Exception {
    return lineNumber;
  }
}
```

Because of the @Named("AttendanceReader"), the batch processing engine knows that this class is meant when the name AttendanceReader is referred to in the job definition file. The injection of the JobContext allows you to address the properties defined in that file. Inside the open() method, you open the input file. If its checkpoint parameter is not null, you wind the input file to the last line defined by the checkpoint info method at the end of the class. Inside readItem(),build and return an AttendanceItem object. Subsequent processor steps get exactly this object.

As such a processor, define a class that filters out items with missing SSNs:

```java
package book.jakartapro.batch;

import jakarta.batch.api.chunk.ItemProcessor;
import jakarta.enterprise.context.Dependent;
import jakarta.inject.Named;

@Dependent
@Named("AttendanceProcessor")
public class AttendanceProcessor implements ItemProcessor {

  @Override
  public Object processItem(Object item) throws Exception {
    // Use this to convert an object, or return null to
    // exclude objects
    AttendanceItem itm = (AttendanceItem) item;
    if(itm.getSsn() == null ||
        itm.getSsn().trim().isEmpty()) return null;
    return item;
  }
}
```

A writer class implements jakarta.batch.api.chunk.ItemWriter and, in its writeItems() method, gets a list of items that are passed. You can use it to create or update a per-employee file summing each employee's workload:

```java
package book.jakartapro.batch;

import java.io.File;
import java.io.FileWriter;
import java.io.Serializable;
import java.util.List;
import java.util.Properties;
import java.util.Scanner;
```

```java
import jakarta.batch.api.chunk.ItemWriter;
import jakarta.batch.runtime.context.JobContext;
import jakarta.enterprise.context.Dependent;
import jakarta.inject.Inject;
import jakarta.inject.Named;

@Dependent
@Named("AttendanceWriter")
public class AttendanceWriter implements ItemWriter {
  private String outputDir;

  @Inject
  private JobContext jobCtx;

  @Override
  public void open(Serializable checkpoint)
        throws Exception {
    Properties props = jobCtx.getProperties();
    outputDir = props.getProperty("output_folder") +
          File.separator + jobCtx.getExecutionId();

    if(checkpoint == null) {
      // first start
      new File(outputDir).mkdirs();
    } else {
      // restart
    }
  }

  @Override
  public void close() throws Exception {
  }

  @Override
  public void writeItems(List<Object> items)
        throws Exception {
    for(Object o : items) {
      AttendanceItem itm = (AttendanceItem) o;
```

```java
      String ssn = itm.getSsn().trim();
      long millis = itm.getWorkMillis();

      File f = new File(outputDir + File.separator + ssn);
      long tm = 0L;
      if(f.exists()) {
        // The following line is just a simple trick
        // to read in the contents of a file.
        try(Scanner s = new Scanner(f).
            useDelimiter("\\Z")) {
          tm = Long.parseLong(s.next().trim());
        }
      }
      tm += millis;

      f.delete();
      FileWriter fw = new FileWriter(f);
      fw.append(tm + "\n");
      fw.flush();
      fw.close();
    }
  }

  @Override
  public Serializable checkpointInfo() throws Exception {
    return null;
  }
}
```

The number of items passed to `writeItems()` corresponds to the `item-count` attribute that's used to define the item's extent between checkpoints.

The last step is a task type step. It backs up the input file and removes empty output folders:

```java
package book.jakartapro.batch;

import java.io.File;
import java.util.Properties;
```

```java
import jakarta.batch.api.AbstractBatchlet;
import jakarta.batch.runtime.context.JobContext;
import jakarta.enterprise.context.Dependent;
import jakarta.inject.Inject;
import jakarta.inject.Named;

@Dependent
@Named("Cleanup")
public class Cleanup extends AbstractBatchlet {

  @Inject
  private JobContext jobCtx;

  @Override
  public String process() throws Exception {
    backupInput();
    removeEmptyOutputDirs();

    return "COMPLETED"; // "FAILED"
  }

  private void removeEmptyOutputDirs() {
    Properties props = jobCtx.getProperties();
    String outputDir = props.getProperty("output_folder") +
          File.separator + jobCtx.getExecutionId();
    if(new File(outputDir).list().length == 0)
      new File(outputDir).delete();
  }

  private void backupInput() {
    Properties props = jobCtx.getProperties();
    String fileName = props.getProperty("input_file");

    String bakName = fileName + ".bak" +
          System.currentTimeMillis();
    new File(fileName).renameTo(new File(bakName));
  }
}
```

Building and Deploying the Application

In order to build and deploy the example application, invoke the Gradle `deployEar` task. You can either do that from inside Eclipse in the Gradle Tasks view (make sure you checked Show All Tasks in the view's menu), or in a terminal using the Gradle wrapper, type the following:

```
cd path/to/root/project
./gradlew deployEar
```

PART IV

Useful Supporting Technologies

CHAPTER 22

XML Binding

XML binding in Java relates to a Java object tree, which is a set of Java objects connected via class member relations, with XML documents. The technology behind the XML binding is called JAXB (Java XML Binding), and for Jakarta EE 10, the JAXB specification is in version number 4.0.

Caution Until Java version 11, the JAXB technology was part of Java SE. Beginning with Java 11, JAXB libraries have to be added explicitly in order for JAXB to work correctly. Since JAXB is part of any Jakarta EE server, you don't have to do this yourself.

Why JAXB Is So Complicated

At first sight, it should be easy to convert an XML document into a Java object tree and vice versa. For example, consider a music record library as an XML document:

```xml
<?xml version="1.0" encoding="UTF-8"?>
<catalog>
  <recordList>
    <record id="173658" type="cd" make="1733">
      <fileType>wav</fileType>
      <genre>classics</genre>
      <composer>J.S. Bach</composer>
      <title>Mass in B-flat Minor</title>
      <performer>H. Masters</performer>
    </record>
    <record id="1734536" type="usb" make="2001">
```

© Peter Späth 2023
P. Späth, *Pro Jakarta EE 10*, https://doi.org/10.1007/978-1-4842-8214-4_22

```
<fileType>mp3</fileType>
<genre>rock</genre>
<composer>G. Jung</composer>
<title>Butterflies</title>
<performer>B. Hoyden</performer>
</record>
</recordList>
</catalog>
```

Here are the Java classes:

```
public class Catalog {
    private List<Record> records;
    // Getters and setters...
}
```

```
public class Record {
    private long id;
    private String typ;
    private int make;
    private String fileType;
    private String genre;
    private String composer;
    private String title;
    private String performer;
    // Getters and setters...
}
```

The challenges for a conversion engine are as follows:

- If you're converting from Java to XML, you must decide which values go as elements, and which values go as attributes. This must be specified.

- From a Java perspective it is possible, although probably not advisable, to add a field to a record that points to the parent: public class Record { ... private Catalog catalog; ... }. A naive conversion to XML in this case would introduce an infinite regression: A record pointing to the catalog, the catalog pointing to all records, each record again pointing to the parent catalog, and so on, forever. There must be a way to avoid this.

- In XML, you would write `<record ... make = "unknown" >`. When converting to the corresponding integer field, this would lead to an error. There must be a way to ensure type safety or impose other pattern constraints to XML values.

- Although not used in this simple example, there must be a common understanding as to how more complex basic types like dates and byte arrays will be converted.

- If names cannot match, you must be able to translate names during conversion. The example uses `type` in XML and `typ` for the corresponding field in Java.

Writing a Java Object Tree as XML

This process, usually called *marshalling*, is handled by Java classes marked with annotations and a JAXB processing engine you get from factories inside the `jakarta.xml.bind` package.

The annotations you can use for XML binding are numerous. You can see the complete list and all detailed descriptions if you look at the `jakarta.xml.bind.annotation` package:

```
https://docs.oracle.com/javase/8/docs/api/
        javax/xml/bind/annotation/package-frame.html
```

For this simple music catalog, use the following annotations:

- `@XmlRootElement`: Add this as an annotation to the class building the root of the object tree. You can also define a namespace for the resulting XML element (including the child hierarchy).

- `@XmlAccessorType`: This annotation allows you to specify how the class defines its data members. For simple cases, you can indicate that fields should be considered for conversions.

- @XmlTransient: Use this annotation to mark fields that have to be skipped for the conversion.

- @XmlElement: If you don't specify field access as @XmlAccessorType, or if you need to explicitly specify an XML element name, you can use this annotation to mark and configure a field for conversion.

- @XmlAttribute: If you want a field to end up as an attribute and not as an XML element, use this annotation at the field level.

- @XmlElementWrapper: For collections, specify the corresponding XML container element using this annotation. Usually, you also add an @XmlElement annotation to specify how the child elements are named.

For this music catalog example, change the Catalog type to the following:

```
package book.jakartapro.xmlbinding;

import java.util.List;
import jakarta.xml.bind.annotation.*;

@XmlRootElement(namespace = "http://book.jakartapro.catalog")
@XmlAccessorType(XmlAccessType.FIELD)
public class Catalog {
  @XmlElementWrapper(name = "recordList")
  @XmlElement(name = "record")
  private List<Record> records;

  // Getters and setters...
}
```

You can see that all fields, here it is just records, will be converted to XML elements. The catalog is marked as the root element, the collection is called recordList, and its elements are called record.

The accordingly adapted Record class reads as follows:

```
package book.jakartapro.xmlbinding;

import jakarta.xml.bind.annotation.*;
```

```
@XmlAccessorType(XmlAccessType.FIELD)
public class Record {
    @XmlAttribute private long id;
    @XmlAttribute(name = "type") private String typ;
    @XmlAttribute private int make;
    private String fileType;
    private String genre;
    private String composer;
    private String title;
    private String performer;

    // Getters and setters...
}
```

This example uses @XmlAttribute to mark fields that you want to go to attributes instead of elements. All the other fields are automatically transformed to XML elements by virtue of the @XmlAccessorType annotation.

For the conversion itself, you use a JAXBContext.newInstance() call and then create a marshaller (exception handling omitted):

```
List<Record> records = new ArrayList<>();

Record r1 = new Record();
r1.setId(173658);
r1.setTyp("cd");
r1.setMake(1733);
r1.setFileType("wav");
r1.setGenre("classics");
r1.setComposer("J.S. Bach");
r1.setTitle("Mass in B-flat Minor");
r1.setPerformer("H. Masters");

Record r2 = new Record();
r2.setId(1734536);
r2.setTyp("usb");
r2.setMake(2001);
r2.setFileType("mp3");
```

```
r2.setGenre("rock");
r2.setComposer("G. Jung");
r2.setTitle("Butterflies");
r2.setPerformer("B. Hoyden");

records.add(r1);
records.add(r2);

Catalog cat = new Catalog();
cat.setRecords(records);

JAXBContext context = JAXBContext.newInstance(
      Catalog.class);
Marshaller m = context.createMarshaller();
m.setProperty(Marshaller.JAXB_FORMATTED_OUTPUT,
      Boolean.TRUE);

// Write to System.out
m.marshal(cat, System.out);

// Write to File
//m.marshal(cat, new File("catalog.xml"));
```

Because of the JAXB_FORMATTED_OUTPUT property, you will get nicely formatted output, and the marshal() method call actually performs the transformation. The output of this sample program looks like this:

```
<?xml version="1.0" encoding="UTF-8" standalone="yes"?>
<ns2:catalog xmlns:ns2="http://book.jakartapro.catalog">
    <recordList>
        <record id="173658" type="cd" make="1733">
            <fileType>wav</fileType>
            <genre>classics</genre>
            <composer>J.S. Bach</composer>
            <title>Mass in B-flat Minor</title>
            <performer>H. Masters</performer>
        </record>
        <record id="1734536" type="usb" make="2001">
            <fileType>mp3</fileType>
```

```
        <genre>rock</genre>
        <composer>G. Jung</composer>
        <title>Butterflies</title>
        <performer>B. Hoyden</performer>
      </record>
    </recordList>
</ns2:catalog>
```

Adding a Schema

In the preceding section, the XML document you retrieved did not have a schema associated with it. XML schemas are not introduced detail here, but I at least want to give you a couple of hints for how to include them.

To include a reference to a schema in the generated XML, add one of the following to the marshaller:

```
m.setProperty(
    Marshaller.JAXB_SCHEMA_LOCATION,
    "http://some.server/catalog.xsd");
// or, if the schema does not specify a namespace
m.setProperty(
    Marshaller.JAXB_NO_NAMESPACE_SCHEMA_LOCATION,
    "http://some.server/catalog.xsd");
```

In order to validate the generated XML against an XSD schema file, you have to first generate a schema file and then add it to the marshaller.

The schema file can be generated manually using the XML Schema specification, or you can use a tool. Eclipse and other IDEs provide schema generators. The one from Eclipse can be accessed via File ➤New ➤ Other... ➤JAXB ➤ Schema from JAXB Classes. Or you can use the Apache tool called XMLBeans for this purpose. For this catalog, a schema file would read as follows:

```
<?xml version="1.0" encoding="UTF-8" standalone="yes"?>
<xs:schema version="1.0"
    xmlns:xs="http://www.w3.org/2001/XMLSchema"
    xmlns:tns="http://book.jakartapro.catalog"
    targetNamespace="http://book.jakartapro.catalog">
```

```xml
<xs:element name="catalog" type="tns:catalogType" />

<xs:complexType name="catalogType">
  <xs:sequence>
    <xs:element name="recordList" minOccurs="0">
      <xs:complexType>
        <xs:sequence>
          <xs:element name="record"
              type="tns:recordType"
              minOccurs="0" maxOccurs="unbounded"/>
        </xs:sequence>
      </xs:complexType>
    </xs:element>
  </xs:sequence>
</xs:complexType>

<xs:complexType name="recordType">
  <xs:sequence>
    <xs:element name="fileType" type="xs:string"
        minOccurs="0"/>
    <xs:element name="genre" type="xs:string"
        minOccurs="0"/>
    <xs:element name="composer" type="xs:string"
        minOccurs="0"/>
    <xs:element name="title" type="xs:string"
        minOccurs="0"/>
    <xs:element name="performer" type="xs:string"
        minOccurs="0"/>
  </xs:sequence>
  <xs:attribute name="id" type="xs:long"
      use="required"/>
  <xs:attribute name="type" type="xs:string"/>
  <xs:attribute name="make" type="xs:int"
      use="required"/>
</xs:complexType>
</xs:schema>
```

To add validation to the code, change the code to read as follows:

```
... [same as above]
records.add(r1);
records.add(r2);

SchemaFactory sf = SchemaFactory.newInstance(
    "http://www.w3.org/2001/XMLSchema");
Schema schem = sf.newSchema(new File("Catalog.xsd"));

Catalog cat = new Catalog();
cat.setRecords(records);

JAXBContext context =
    JAXBContext.newInstance(Catalog.class);
Marshaller m = context.createMarshaller();
m.setSchema(schem);   // <==== NEW! ====
... [same as above]
```

Of course, ... new File() ... must be changed to point to your schema file. If you start the application, the generated XML will be validated against the schema file you provided.

Note   For this to work in a non-Jakarta EE environment, a JAXP implementation must be added to the classpath. One candidate is the Xerces library, for example, added via the Maven dependency artifact = "xercesImpl", group = "xerces", version = "2.12.1".

Transforming from XML to Java

For the opposite operation, transforming XML to a Java object tree, the procedure is very similar. You use JAXBContext and create an unmarshaller from it. For the music records example, this process reads as follows:

```
String xml = """
  <?xml version=\"1.0\" encoding=\"UTF-8\"
      standalone=\"yes\"?>
  <ns2:catalog
    xmlns:ns2=\"http://book.jakartapro.catalog\"
    xmlns:xsi=" +
        \"http://www.w3.org/2001/XMLSchema-instance\"
    xsi:schemaLocation=
        \"http://book.jakartapro.catalog\">
      <recordList>
          <record id=\"173658\" type=\"cd\"
                make=\"1733\">
              <fileType>wav</fileType>
              <genre>classics</genre>
              <composer>J.S. Bach</composer>
              <title>Mass in B-flat Minor</title>
              <performer>H. Masters</performer>
          </record>
          <record id=\"1734536\" type=\"usb\"
                make=\"2001\">
              <fileType>mp3</fileType>
              <genre>rock</genre>
              <composer>G. Jung</composer>
              <title>Butterflies</title>
              <performer>B. Hoyden</performer>
          </record>
      </recordList>
  </ns2:catalog>
  """;
```

```
JAXBContext context = JAXBContext.
    newInstance(Catalog.class);
Unmarshaller um = context.createUnmarshaller();

ByteArrayInputStream bis =
    new ByteArrayInputStream(xml.getBytes());
Object o = um.unmarshal(bis);
System.out.println(o.getClass());
Catalog c = (Catalog) o;
System.out.println(c.getRecords().size());
System.out.println(c.getRecords().get(0).getComposer());
```

In order to add a schema validation, all you have to do is write the following:

```
SchemaFactory sf = SchemaFactory
    .newInstance("http://www.w3.org/2001/XMLSchema");
Schema schem = sf.newSchema(new File("NewXMLSchema2.xsd"));
•••
JAXBContext context = JAXBContext.
    newInstance(Catalog.class);
Unmarshaller um = context.createUnmarshaller();
um.setSchema(schem);
•••
```

Again, use a version of .newSchema() that suits your needs.

Generating Java Classes from Schemas

If you start development given an XSD file, and you want Java classes to be generated automatically, you can use IDE functionalities. In Eclipse, choose File ➤ New ➤ Other... ➤ JAXB ➤ Schema from JAXB Classes. Or you can use a command-line utility like scomp from the Apache XMLBeans project.

The Eclipse XML/XSD tools seem to have problems with some JRE versions. Therefore, take a closer look at the Apache XMLBeans toolset. Download XMLBeans from `https://xmlbeans.apache.org`. Then run the following in a shell:

```
# Make bin files executable. You have to run this only
# once, after installation
cd /apache/xmlbeans/inst/folder
chmod a+x bin/*

# Set some paths
# (tested with OpenJDK 11)
export JAVA_HOME=/path/to/jdk
export export XMLBEANS_HOME=/path/to/xmlbeans/installation

# Perform the generation
# Afterwards, find Java classes in the src folder
bin/scomp -srconly -src src -d xsb \
  /path/to/xsd/file
```

For details and options, consult the Apache XMLBeans documentation.

JSON Handling

The XML binding technology covered in Chapter 22 introduced a valuable data
exchange format, simplifying and standardizing the communication in heterogeneous
and wide-spread computer networks. By using schemas and namespaces for elements
and attributes, this enabled semantics and version consistency in XML, thus allowing for
XML document validity checks. Unfortunately, with all those additions, XML documents
became more and more complex, and in order to send a few bytes formatted as XML,
a lot of boilerplate data had to be transmitted, too. Especially with browser-to-server
communication, XML became overkill. For this reason, web application developers
switched to the leaner JSON (JavaScript Object Notation) data protocol, making it
necessary for the Java world to develop tools and libraries that could handle JSON.

JSON processing is part of the Jakarta EE 10 specification. This specifically involves
Jakarta JSON Binding (JSON-B), version 3.0, and Jakarta JSON Processing (JSON-P),
version 2.1. The specifications for JSON-P and JSON-B can be found at `https://jcp.`
`org/aboutJava/communityprocess/pr/jsr367/index.html` and `https://jcp.org/en/`
`jsr/detail?id=374`.

This chapter covers these JSON libraries and explains how you can add JSON to
Jakarta EE applications.

JSON Documents

As is the case for XML, a JSON document represents a tree-like object hierarchy. Unlike
XML, JSON entities cannot have attributes. There are no such things as namespaces and
schemas, and instead of opening and closing tags, complex JSON objects use { ... } to
map name-value pairs and [...] for arrays. Here is a simple JSON document showing
all features:

```
{
    "lastname": "Smith",
```

P. Späth, *Pro Jakarta EE 10*, https://doi.org/10.1007/978-1-4842-8214-4_23

```
        "married": true,
        "birthday": "1997-12-30",
        "records": null,
        "height": 6.1,
        "cars": 2,
        "children": [
            "Amanda",
            { "forename": "Jim", "gender": "m" }
        ]
        "address": {
            "us": true,
            "state": "California",
            "city": "Paradise",
            "owned": false
        }
    }
```

You can see that basic types are numbers with and without fraction digits, boolean literals true and false, strings, and null. Arrays and maps can contain elements of mixed types, including inner complex types.

JSON with REST Services

In previous chapters, you saw a REST method that returned JSON data, as follows:

```
@GET
@Produces("application/json")
public String stdDate() {
    return "{\"date\":\"" +
        ZonedDateTime.now().toString() + "\"}";
}
```

While at first sight, this seems to be an easy way to create JSON data, it lacks a basic syntax check. First, it easily might happen that you forget a closing }, or that you make a mistake because of misplaced quotation and or punctuation mark. Second, think of a more complex JSON document with more elements and deeper hierarchy. Manually constructing concatenating strings to represent larger JSON objects is a nightmare in terms of comprehensiveness and maintainability.

The first steps toward increased readability consists of using JSON-P to construct JSON data. JSON-P contains classes representing values, arrays, and maps. Replacing the previous code snippet, you will eventually get code like this:

```
...
import jakarta.json.Json;
import jakarta.json.JsonStructure;
...
@GET
@Path("j")
@Produces("application/json")
public JsonStructure stdDate() {
    JsonStructure json = Json.createObjectBuilder()
        .add("date", ZonedDateTime.now().toString())
    .build();
    return json;
}
```

You can see that you no longer have to concatenate strings to build the JSON document. You can even let the method return an object of type JsonObject, JsonArray or JsonStructure (all from the jakarta.json package). JAX-RS knows how to handle such return types.

From AJAX calls and if you need to pass data from the web frontend to the server, you can either let the REST service receive a string, as in this:

```
// Corresponds to for example AJAX calls from jQuery:
// $.ajax({
//   method: "POST",
//   contentType: "application/json",
//   url: "../webapi/date",
//   data: { timeZone: "UTC", format: "standard" }
// })
// .done(function( msg, textStatus, jqXHR ) {
//   alert( "Data Saved: " + msg );
// });
```

```
@POST
@Path("/date")
@Consumes(MediaType.APPLICATION_JSON)
...
public Response stdDate(String data) {
    // data is a JSON string
    ...
}
```

Or, as an alternative, you can use a jakarta.json.JsonStructure parameter for the method, as follows:

```
// Corresponds to for example AJAX calls from jQuery:
// $.ajax({
//   method: "POST",
//   contentType: "application/json",
//   url: "../webapi/date",
//   data: '{ timeZone: "UTC", format: "standard" }'
// })
// .done(function( msg, textStatus, jqXHR ) {
//   alert( "Data Saved: " + msg );
// });

@POST
@Path("/date")
@Consumes(MediaType.APPLICATION_JSON)
...
public Response stdDate(JsonStructure data) {
    ...
}
```

For the jQuery AJAX call, you must pass a JSON formatted string, not the JavaScript data object.

Apart from these simple examples, there is more to discover about JSON handling using the JSON-P and JSON-B technologies. The following sections more thoroughly explain JSON binding and processing.

Generating JSON

In order to generate JSON data using the builder pattern, you use the classes from the jakarta.json package. You start from a builder:

```
...
import jakarta.json.*;
...
    JsonObjectBuilder builder = Json.createObjectBuilder();
```

From here you add elements as follows:

```
builder.add("elementName", ...);
// <- you can add int, long, double, boolean,
//    String, BigInteger, and BigDecimal elements.
//    And you can append more add(...) statements.
```

There is also a variant for null values and for complex types:

```
builder.addNull("elementName");

builder.
    add("element1",
        Json.createObjectBuilder().
            add("inner1", 42).
            add("inner2", "deepThought")).
    add("element2", 3.1415);
```

To create JSON arrays, use the JsonArrayBuilder class instead:

```
...
import jakarta.json.*;
...
    JsonArrayBuilder builder = Json.createArrayBuilder();
    builder.add("Fibonacci");
    builder.add(1); // it is allowed to mix types in arrays
    builder.add(2);
    builder.add(3);
    builder.add(5);
```

```
builder.add(8);
```

For both JSON objects and JSON arrays, to get the Java object, you must invoke the build() method:

```
...
import jakarta.json.*;
...
    JsonObjectBuilder builder = Json.createObjectBuilder();
    // ... add elements, then:
    JsonObject obj = builder.build();
...
    JsonArrayBuilder abuilder = Json.createArrayBuilder();
    // ... add elements, then:
    JsonArray aobj = abuilder.build();
```

Parsing JSON

For low-level parsing of JSON strings, you can use JSON-P's streaming API:

```
...
import jakarta.json.*;
...
    final String json =
      "{\"name\":\"John\",\"age\":27,\"employed\":false}";
    final JsonParser parser =
      Json.createParser(new StringReader(json));
    String key = null;
    String value = null;
    while (parser.hasNext()) {
        final Event event = parser.next();
        switch (event) {
        case KEY_NAME:
            key = parser.getString();
            System.out.println(key);
            break;
```

CHAPTER 23   JSON HANDLING

```
        case VALUE_STRING:
            value = parser.getString();
            System.out.println(value);
            break;
        }
    }
    parser.close();
```

This comes in handy if you don't need a verbose JSON-to-Java object binding, which requires extra coding. Also, if you want to read in a very large JSON document, the streaming API helps you avoid allocating exhaustive resources.

If you don't like the pull-parsing API, you can read in a JSON string, thus producing a generic JSON object instead:

```
final String json =
    "{\"name\":\"John\",\"age\":27,\"employed\":false}";

JsonReader jsonReader =
    Json.createReader(new StringReader(json));

JsonObject object = jsonReader.readObject();
jsonReader.close();
...
String name = object.getString("name", "<unknown>");
```

For more details, see the API documentation at https://javaee.github.io/jsonp/index.html.

Binding JSON to Java Objects

If you need an automatic mapping between a Java object tree and a JSON representation of such a tree, you can use the classes from the JSON-B library. Consider the following JSON string for a club members list:

```
[
  {
    "name": "John",
    "age": 27,
```

```
    "employed":false,
    "children" :
      [
        {"name": "Bart", "birthday": "2015-11-10"},
        {"name": "Amanda", "birthday": "2017-06-01"}
      ]
  },
  {
    "name": "Linda",
    "age": 22,
    "employed":true,
    "children" :
      [
      ]
  }
]
```

A natural set of classes representing the same list would be as follows:

```
class Member {
  private String name;
  private int age;
  private boolean employed;
  private List<Child> children;
  // getters and setters...
}

class Child {
  private String name;
  private String birthday;
  // getters and setters...
}
```

In order to read in the JSON string and construct a list of members, all you have to do is this:

```
String json = "[ ... ]";
```

```
Jsonb jsonb = JsonbBuilder.create();

@SuppressWarnings("unchecked")
List<Member> members = jsonb.fromJson(json, List.class);

// just to check, the other way round, getting the
// JSON string from a Java object tree:
jsonb = JsonbBuilder.create();
String json2 = jsonb.toJson(members);
// <- semantically the same as 'json' above.
```

Converting between simple types like strings, booleans, and numbers is straightforward. If you need more control over the conversion process, JSON-B helps you with a couple of annotations. In the previous example, you can for example change the birthday field to a field of type `java.util.Date`:

```
class Child {
  private java.util.Date birthday;
  ...
}
```

(The getters and setters are accordingly adapted.) The `jakarta.json.bind.annotation.JsonbDateFormat` then allows for the specification of a date string conversion pattern. For the `yyyy-MM-dd` pattern, see the API documentation of `SimpleDateFormat` for more options. The annotation would read as follows:

```
class Child {
  @JsonbDateFormat("yyyy-MM-dd")
  private java.util.Date birthday;
  ...
}
```

More annotations control the conversion of number formats and `null` values. It is also possible to exclude fields from conversion, and for maximum control, you can define a custom type adapter. Table 23-1 summarizes the JSON-B annotations.

Table 23-1. *JSON-B Annotations*

Annotation	Useful For
@JsonbDateFormat	Defining a custom date format. Mediates between a java.util.Date in Java and a JSON string value.
@JsonbNillable	Showing a null Java value as "null" in the JSON string. If set to false, a null value does not show up in the JSON string.
@JsonbNumberFormat	Defining a custom number format. Mediates between a java.math. BigDecimal in Java and a JSON string value. Format options are described in the API documentation of the DecimalFormat class.
@JsonbTransient	Excluding a field from conversion.
@JsonbProperty	Specifying a custom key name for a mapping.
@JsonbTypeAdapter	Specifying a custom type adapter (the JsonbAdapter class).

As an example of a type adapter, consider the following classes:

```java
class Color {
  public int r;
  public int g;
  public int b;
  public Color(int r, int g, int b) {
    this.r = r;
    this.g = g;
    this.b = b;
  }
}

class Shape {
  private Color color;
  ...
  // getters and setters...
}
```

If you want a color to be representable as the hexadecimal value #RRGGBB, a JSON snippet for a shape object could look like this:

```
{
  "color": "#FF1B75",
  ...
}
```

The converter class is responsible for translating the Color object to a JSON string representation and vice versa:

```java
class MyColorAdapter implements
      JsonbAdapter<Color, String> {
  @Override
  public String adaptToJson(Color obj)
  throws Exception {
    return "#" +
      String.format("%02x", obj.r) +
      String.format("%02x", obj.g) +
      String.format("%02x", obj.b);
  }

  @Override
  public Color adaptFromJson(String obj)
  throws Exception {
    int r = Integer.parseInt( obj.substring(1,3), 16);
    int g = Integer.parseInt( obj.substring(3,5), 16);
    int b = Integer.parseInt( obj.substring(5,7), 16);
    return new Color(r,g,b);
  }
}
```

Finally, to add the converter to the Shape class, you have to apply the @JsonbTypeAdapter annotation as follows:

```java
class Shape {
  @JsonbTypeAdapter(MyColorAdapter.class)
  private Color color;
  ...
  // getters and setters...
}
```

The documentation of the annotations from the `jakarta.json.bind.annotation` package shows more information. To learn more about JSON-B, check out `https://javaee.github.io/jsonb-spec/` as well.

CHAPTER 24

Jakarta Mail

Being able to send emails from server processes can be a vital part of a company's IT landscape. Consider, for example, asynchronously transmitting results from long-running calculations, sending automatically generated reports, or sending warning messages when a server process fails.

Not often recognized as a central part of the Jakarta EE technology stack, Jakarta Mail 2.1 can serve this purpose. This chapter covers Jakarta Mail, from `https://eclipse-ee4j.github.io/mail/`, which is included with Jakarta EE 10 as a versatile email library.

Installing Jakarta Mail

Jakarta Mail is included in the Jakarta EE specification, so for any enterprise project with this basic dependency (Gradle configuration, Maven accordingly):

```
implementation 'jakarta.platform:' +
    'jakarta.jakartaee-api:10.0.0'
```

You have everything you need to use Jakarta Mail.

Note Jakarta Mail's predecessor is called JavaMail. The main difference is that Jakarta Mail uses the `jakarta.` namespace instead of `javax.` Many online resources still refer to JavaMail. In most cases, you can easily adapt JavaMail tutorials by simply switching to the new namespace.

Generating and Sending Emails

The standard way to send emails from a Jakarta EE application consists of the following:

331

© Peter Späth 2023
P. Späth, *Pro Jakarta EE 10*, https://doi.org/10.1007/978-1-4842-8214-4_24

- Provide an authenticator for authentication at the Message Transfer Agent (MTA). To this aim, create an instantiable class that inherits from jakarta.mail.Autenticator. If you don't need to authenticate to your MTA, you can skip this step.

- Instantiate and fill a java.util.Properties collection. Here, various SMTP-related parameters need to be entered.

- Get a jakarta.mail.Session instance.

- Instantiate a message object, for example jakarta.mail.internet. MimeMessage.

- Get a transport object (the jakarta.mail.Transport class) from the session.

- Use the transport object to connect to the MTA and then send the message.

This example provides a rudimentary REST class for building and sending emails. This is a starting point only. For your own application, you should read the constants from the configuration file. You also have to update the connection parameters to match your mail server.

```java
import java.util.Date;
import java.util.Properties;

import jakarta.mail.Authenticator;
import jakarta.mail.Message;
import jakarta.mail.MessagingException;
import jakarta.mail.PasswordAuthentication;
import jakarta.mail.Session;
import jakarta.mail.Transport;
import jakarta.mail.internet.InternetAddress;
import jakarta.mail.internet.MimeMessage;
import jakarta.ws.rs.POST;
import jakarta.ws.rs.Path;
import jakarta.ws.rs.Produces;

@Path("/mail")
public class JakartaMail {
    final String MAILSERVER_ADDR = "addr.to.mta.com";
```

```
final String MAILSERVER_PORT = "465"; //465,587;
final String MAILSERVER_IDENTITY = "your@email.addr.com";
final String MAILSERVER_LOGIN = "mailserver.login";
final String MAILSERVER_PASSWD = "mailserver.passwd";

@POST
@Produces("application/json")
public String sendMail() {
  sendMail("recipient@example.com", "test",
      new Date().toString());
  return "{\"done\":true}";
}

class SMTPAuthenticator extends Authenticator {
  public PasswordAuthentication
  getPasswordAuthentication(){
    return new PasswordAuthentication(MAILSERVER_LOGIN,
        MAILSERVER_PASSWD);
  }
}

private void sendMail(String toAddr,
      String theSubject,
      String theMessage) {
  Properties props = new Properties();

  props.put("mail.smtp.user", MAILSERVER_IDENTITY);
  props.put("mail.smtp.host", MAILSERVER_ADDR);
  props.put("mail.smtp.port", MAILSERVER_PORT);
  props.put("mail.smtp.starttls.enable","true");
  props.put("mail.smtp.debug", "true");
  props.put("mail.smtp.auth", "true");
  props.put("mail.smtp.socketFactory.port",
      MAILSERVER_PORT);

  //props.put("mail.smtps.ssl.checkserveridentity",
  // "false");
  props.put("mail.smtps.ssl.trust",  MAILSERVER_ADDR);
```

```
SMTPAuthenticator auth = new SMTPAuthenticator();
Session session = Session.getInstance(props, auth);
session.setDebug(false);

MimeMessage msg = new MimeMessage(session);
try {
  msg.setText(theMessage);
  msg.setSubject(theSubject);
  msg.setFrom(new InternetAddress(
      MAILSERVER_IDENTITY));
  msg.addRecipient(Message.RecipientType.TO,
      new InternetAddress(toAddr));

  // use "smtps" instead of "smtp", to enable SSL
  Transport transport = session.getTransport("smtps");
  transport.connect(MAILSERVER_ADDR,
      Integer.parseInt(MAILSERVER_PORT),
      MAILSERVER_LOGIN, MAILSERVER_PASSWD);
  transport.sendMessage(msg, msg.getAllRecipients());
  transport.close();
} catch (MessagingException e) {
  e.printStackTrace(System.err);
  }
 }
}
```

A word about the `mail.smtps.ssl.trust` property: If it's used as shown here, the handshake process used with SSL encrypted communication disregards the server's SSL certificate check. If you comment out this line, you might have to import the server's certificate into Java's `cacerts` key store. So expect some additional work if you need to check the server's identity.

Once it's deployed—for example as `JakartaMail.war` on a GlassFish server and with the `<url-pattern>` /webapi/* `</url-pattern>` URL pattern in `web.xml`—you can trigger an email from a terminal via this command:

```
curl -X POST \
http://localhost:8080/JakartaMail/webapi/email
```

## CHAPTER 25

Application Client (Groovy)

Jakarta EE server applications usually provide interfaces that clients can use to fulfill enterprise use case needs. These are often web or REST services for communicating XML or JSON data, or EJB interfaces that a client can access via RMI. However, during the development phase and later for maintenance purposes like monitoring, a lightweight and easy-to-configure access library is better solution. Some application servers like GlassFish therefore provide scripts for generating such on-the-fly client libraries, commonly referred to as *application clients*.

Note Application clients are not standardized. For Jakarta EE servers other than GlassFish, you have to consult the documentation to see whether application clients are supported. Or you can develop your own scripts.

This chapter is about using Groovy scripts to access EJB interfaces via an application client library generated by GlassFish.

Providing an Enterprise Application

You can start from any Jakarta EE application with a `remote` EJB interface. A remote interface is necessary, since you run the client code in a JVM that's different from the application server.

This simple example codes an EAR project named `ApplicationClientServer`. It targets a `ApplicationClientServer.ear` deployable and uses the following EJB classes:

```
// ----- file TheDate.java
```

© Peter Späth 2023
P. Späth, *Pro Jakarta EE 10*, https://doi.org/10.1007/978-1-4842-8214-4_25

```
package book.jakartapro.appclient.ejb;

import java.time.ZonedDateTime;

import book.jakartapro.appclient.ejb.interfaces.
    TheDateLocal;
import book.jakartapro.appclient.ejb.interfaces.
    TheDateRemote;
import jakarta.ejb.Local;
import jakarta.ejb.Remote;
import jakarta.ejb.Stateless;

@Stateless
@Local(TheDateLocal.class)
@Remote(TheDateRemote.class)
public class TheDate {
    public String fetchDate() {
        return ZonedDateTime.now().toString();
    }
}

// ----- file TheDateLocal.java
package book.jakartapro.appclient.ejb.interfaces;

public interface TheDateLocal {
    String fetchDate();
}

// ----- file TheDateRemote.java
package book.jakartapro.appclient.ejb.interfaces;

public interface TheDateRemote {
    String fetchDate();
}
```

> **Note**   At this point, the example doesn't explain how to develop a Jakarta EE application packaged as an EAR file. Different IDEs use different workflows for this aim, and it also depends on which build framework you use. Previous chapters provided hints; consult your server documentation if you need more information.

For the application client generation to work, this EAR must be deployed and running on GlassFish.

Building an Application Client

With an EAR application deployed on GlassFish, the steps for generating and running application clients are summarized as follows:

1. Package the interface classes into a `ejb-interfaces.jar` archive.

2. Generate `appclient.jar`. Go to the `GLASSFISH_INST/glassfish` folder and issue a `bin/package-appclient`.

3. Fetch the `GLASSFISH_INST/glassfish/lib/appclient.jar` generated in the previous step.

4. Unzip `appclient.jar` to any suitable folder on the machine (remember a `.jar` file is just a ZIP file with a different suffix). Call this folder `MYAPPCLIENT`. Move the contents of the `appclient` folder to its parent folder.

5. Copy `ejb-interfaces.jar` into `MYAPPCLIENT`.

6. Modify the `MYAPPCLIENT/config/asenv.*` and `MYAPPCLIENT/glassfish/domains/domain1/config/sun-acc.xml` files. See the instructions at `https://glassfish.org/docs/latest/reference-manual.html#package-appclient`. This step is necessary only if you want to use the GlassFish client application container. This example use a different approach; see the next section.

7. Make the MYAPPCLIENT/glassfish/bin/appclient and
    MYAPPCLIENT/glassfish/bin/appclient.bat files executable.
    This step is necessary only if you want to use the GlassFish client
    application container. This example uses a different approach; see
    the next section.

8. If you want to start the application client container, provide
    the client code as, for example, myClient.jar. Then run
    MYAPPCLIENT/glassfish/bin/appclient, using myClient.jar as
    a parameter.

Running a GroovyConsole Application Client

To allow a Groovy script to access the Jakarta EE server, first download and install Groovy
from https://groovy-lang.org/install.html. Next—with GROOVY_HOME pointing to
the Groovy installation folder and JAVA_HOME pointing to JDK version 11, and with the
MYAPPCLIENT folder prepared as described in the preceding section—build a shell script
inside MYAPPCLIENT as follows:

```
#!/bin/bash
# ---------- File 'groovyForJakarta.sh' ----------

# Provide your own paths here:
export GROOVY_HOME=/opt/groovy-4
export JAVA_HOME=/opt/jdk17

addClasspath="ejb-interfaces.jar"
for i in glassfish/modules/*.jar; do
  addClasspath=${addClasspath}:$i
done

export GROOVY_TURN_OFF_JAVA_WARNINGS=true

$GROOVY_HOME/bin/groovyConsole -cp $addClasspath
```

Unfortunately, there is a bug in the startup script of some versions of the Groovy
distribution. Check the GROOVY/bin/startGroovy file for this line:

```
if [ "$(expr "$JAVA_VERSION" \> "1.8.0")" ]; then
```

and if it exists, replace it by

```
if [ "$(expr true)" ]; then
```

Obviously 1.17.x is not greater than 1.8.0, so the original condition is wrong.

When you start the script, a GroovyConsole window appears. To access EJBs from that window, you can use code like this:

```
import javax.naming.*

def ctx = getContext()

// This is GlassFish' JNDI naming schema
def NAMING_SCHEMA = "java:global"

// The name of the application. For GlassFish, the name
// of the .ear file, without the suffix
def APPL_NAME = "AppClientServer"

// The EJB module name inside the EAR file
def EJB_MODULE_NAME = "AppClientEjb"

// The EJB's class name
def EJB_NAME = "TheDate"

// The fully qualified name of the EJB's remote interface
def EJB_REMOTE_CLASS =
  "book.jakartapro.appclient.ejb.interfaces.TheDateRemote"

def look = "${NAMING_SCHEMA}/" +
  "${APPL_NAME}/${EJB_MODULE_NAME}/" +
  "${EJB_NAME}!${EJB_REMOTE_CLASS}"
def ejb = ctx.lookup(look)

def date = ejb.fetchDate() // invoke the ejb
println date

def getContext() {
  Properties props = new Properties().with(true){
    setProperty(Context.INITIAL_CONTEXT_FACTORY,
      "com.sun.enterprise.naming.SerialInitContextFactory")
```

```
    setProperty("org.omg.CORBA.ORBInitialHost",
      "localhost") // application server
    setProperty("org.omg.CORBA.ORBInitialPort",
      "3700") // Glassfish' JNDI port
  }
  new InitialContext(props)
}
```

## CHAPTER 26

Adding Scripting Languages

The Java language is not always the best choice for implementing business functionalities and algorithms. The language only slowly implements new features, which can be an advantage, because it is easier to maintain old code when the coding structures don't change too often. And Java is statically typed, which means you have to take care of object types at a language level. This leads to more typing work, since type specifiers need to be entered often in order to make sure that type conformance works correctly. Contrary to that, scripting languages like Groovy and Python often include features to write code more concisely, and they are often typed dynamically. That means that type checks don't take place during compiling and the code contains fewer boilerplate constructs to ensure variable type conformance in assignments.

Fortunately, Java allows you to mix Java language code and scripting languages by means of JSR 223. The Java code is compiled before you have any deployable (and runnable) artifacts like class files and class file archives, while the scripts are compiled at runtime. This opens another usage scenario for scripts: Because they can be changed without having to compile the application over and over again, scripts can be changed more easily even after an application enters the production stage. Scripts can, for example, be saved as text in a database, or in resource files extraneous to deployment artifacts, like JARs, WARs, and EARs.

This chapter explains how to include Groovy and Python scripts in a Jakarta EE application.

Installing Scripting Engines

In order to use JSR 223 scripting, you have to add the following dependencies to your project:

© Peter Späth 2023
P. Späth, *Pro Jakarta EE 10*, https://doi.org/10.1007/978-1-4842-8214-4_26

```
// For Python, in Gradle build.gradle file
implementation 'org.python:jython-standalone:2.7.2'

<!-- For Python, in Maven pom.xml file -->
<dependency>
    <groupId>org.python</groupId>
    <artifactId>jython-standalone</artifactId>
    <version>2.7.2</version>
</dependency>

// For Groovy, in Gradle build.gradle file
implementation 'org.apache.groovy:groovy-jsr223:4.0.8'

<!-- For Groovy, in Maven pom.xml file -->
<dependency>
    <groupId>org.apache.groovy</groupId>
    <artifactId>groovy-jsr223</artifactId>
    <version>4.0.8</version>
</dependency>
```

Using Scripting in Java

In order to use a script from anywhere inside your code, write this code (Python version):

```
import javax.script.ScriptEngine;
import javax.script.ScriptEngineManager;
import javax.script.ScriptException;
...
    ScriptEngine engine = new ScriptEngineManager().
        getEngineByName("python");

    // read in Python code
    StringReader f = new StringReader(
            "class Multiplier:\n"
        + "\n"
        + "  def multiply(self, x, y):\n"
        + "    return x * y\n"
```

```
                + "\n"
                + "x = Multiplier().multiply(par1, par2)\n"
    );
    // <- could also use a FileReader here

    try {
        // set input parameters
        engine.put("par1", 5);
        engine.put("par2", 8);
        // evaluate the script
        engine.eval(f);
    } catch (ScriptException e) {
        // ... handle exception
    }
    // fetch output variable
    Object x = engine.get("x"); // -> 40
...
```

The Groovy variant for the same script reads as follows:

```
import javax.script.ScriptEngine;
import javax.script.ScriptEngineManager;
import javax.script.ScriptException;
...
    ScriptEngine engine = new ScriptEngineManager().
        getEngineByName("groovy");

    // read in Groovy code
    StringReader f = new StringReader(
        "class Multiplier {\n"
        + "  def multiply(def x, def y) {\n"
        + "    x * y\n"
        + "  }\n"
        + "}\n"
        + "x = new Multiplier().multiply(par1, par2)\n"
    );
    // <- could also use a FileReader here
```

```java
try {
    // set input parameters
    engine.put("par1", 5);
    engine.put("par2", 8);
    engine.eval(f);
} catch (ScriptException e) {
    // ... handle exception
}
// fetch output variable
Object x = engine.get("x"); // -> 40
```

PART V

Advanced Resource Related Topics

## CHAPTER 27

Hibernate as ORM

The GlassFish Jakarta EE application server comes with a built-in EclipseLink as a JPA provider, and other application servers have their own ideas about which JPA provider to use. After all, it's one of the strengths of JPA that application developers don't have to worry about which JPA provider is used. If you do want to prescribe the JPA provider for some reason, you can do so without having to change the application server installation.

   This chapter explains how to configure the application to use Hibernate as a JPA provider. The procedure is similar when using other providers.

Note   Go to `https://hibernate.org/` to learn what can be done with Hibernate.

Installing Hibernate

To install Hibernate as a JPA provider, add the following code to your `build.gradle` or `pom.xml` file, for Gradle or Maven, respectively:

```
...
dependencies {
    ...
    implementation 'org.hibernate:' +
        'hibernate-core-jakarta:5.6.3.Final'
}
  ...
  <dependencies>
    ...
    <dependency>
```

© Peter Späth 2023
P. Späth, *Pro Jakarta EE 10*, https://doi.org/10.1007/978-1-4842-8214-4_27

```
      <groupId>org.hibernate</groupId>
      <artifactId>hibernate-core-jakarta</artifactId>
      <version>5.6.3.Final</version>
    </dependency>
  </dependencies>
...
```

Adapting the Persistence Configuration

You need to tell the JPA configuration to use Maven as a JPA provider. To this aim, open persistence.xml inside the META_INF folder and add the following:

```
<persistence ...>
  <persistence-unit name="...">
    <provider>
      org.hibernate.jpa.HibernatePersistenceProvider
    </provider>
    <jta-data-source>...</jta-data-source>
    <properties>
        ...
<!-- Specify the database dialect. "DerbyTenSevenDialect"
    is for the Apache Derby DBMS. See the Hibernate
    documentation for other dialects. -->
<property name="hibernate.dialect"
    value="org.hibernate.dialect.DerbyTenSevenDialect" />
<!-- More Hibernate properties. The following might be
    necessary if you have problems with the JTA
    (transactions) manager. "SunOneJtaPlatform" is
    for the Apache Derby database - for other DBMS',
    watch out for the other classes in the
      org.hibernate.engine.transaction.jta.platform.
      internal
    package. -->
<!-- <property
  name = "hibernate.transaction.jta.platform"
```

```
value ="org.hibernate.engine.transaction.jta.platform.
        internal.SunOneJtaPlatform"/>
-->
      </properties>
   </persistence-unit
</persistence>
```

The Hibernate documentation tells you more about the Hibernate settings you can place in this file.

Fetching the Hibernate Session

To directly address the Hibernate JPA given an injected entity manager, the API documentation suggests using the following code:

```
...
import org.hibernate.Session;
...
    @PersistenceContext(unitName = "my-persistence-unit")
    private EntityManager entityManager;
...
    Session session = entityManager.unwrap(Session.class);
    // directly use the Hibernate session:
    // ...
```

Caution   Unfortunately, due to class-loading issues, this does not currently work for GlassFish 7.

## CHAPTER 28

Connectors

It might not be possible for a Jakarta EE application to seamlessly communicate with legacy EISs (enterprise information systems) using one of the standard communication protocols. This is especially true in corporate environments. For this to be possible, you have to create a resource adapter that conforms to the Jakarta Connectors architecture.

The Jakarta Connectors technology (version number 2.1) is part of the Jakarta EE 10 technology stack. The JCA project's home page can be found at `https://projects.eclipse.org/projects/ee4j.jca/releases/2.1.0`, and for the specification, see `https://www.jcp.org/en/jsr/detail?id=322`.

Creating a connector for a big or important EIS can be a very challenging task. The specification document has more than 500 pages, so this book covers only a few features of the Connectors architecture. This chapter works through a simple JCA project, which nevertheless should give you a starting point for your own projects.

Coding a Resource Adapter

For a resource adapter, you need at least the following artifacts:

- An interface describing all the business methods of the resource adapter, accessible from a client. Usually referred to as *connections*. In this example, it's called `HelloWorldConnection`.

- Another interface describing a connection factory. Its sole purpose is providing a factory method for obtaining connections. It must extend `jakarta.resource.Referenceable` and `java.io.Serializable`. In this example, it's called `HelloWorldConnectionFactory`.

- A class implementing the connection interface. In this example, it's called `HelloWorldConnectionImpl`.

© Peter Späth 2023
P. Späth, *Pro Jakarta EE 10*, https://doi.org/10.1007/978-1-4842-8214-4_28

- A class implementing the connection factory interface. In this example, it's called `HelloWorldConnectionFactoryImpl`.

- A class extending `jakarta.resource.spi.ManagedConnection`. This represents a physical connection to the underlying EIS. In this example, it's called `HelloWorldManagedConnection`.

- A factory class for a `ManagedConnection`. It must implement `jakarta.resource.spi.ManagedConnectionFactory` and `jakarta.resource.spi.ResourceAdapterAssociation`. In this example, it's called `HelloWorldManagedConnectionFactory`.

- A class implementing `jakarta.resource.spi.ManagedConnectionMetaData` for metadata. In this example, it's called `HelloWorldManagedConnectionMetaData`.

- A class implementing `jakarta.resource.spi.ResourceAdapter`. In this example, it's called `HelloWorldResourceAdapter`.

For more details about the class relations and descriptions of what each class does, consult the API documentation and the JCA specification.

To simplify packaging, this example puts all the interfaces into a `interfaces` subpackage. Figure 28-1 shows the project layout for this example project.

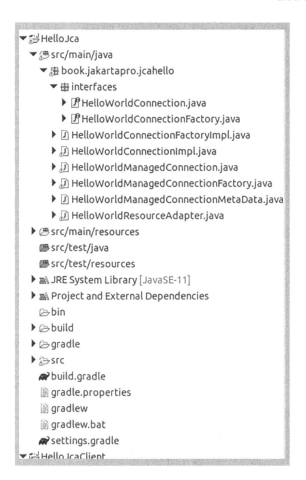

Figure 28-1. *JCA example project layout*

The coding for the two interfaces is as follows:

```java
// ---------- File HelloWorldConnection.java
package book.jakartapro.jcahello.interfaces;

public interface HelloWorldConnection {
    public String helloWorld();
    public String helloWorld(String name);
    public void close();
}

// ---------- File HelloWorldConnection.java
package book.jakartapro.jcahello.interfaces;
```

```java
import java.io.Serializable;

import jakarta.resource.Referenceable;
import jakarta.resource.ResourceException;

public interface HelloWorldConnectionFactory extends
      Serializable, Referenceable {
   public HelloWorldConnection getConnection()
      throws ResourceException;
}
```

The two classes implementing these interfaces are shown next. Note how the helloWorld() method accesses a resource adapter property.

```java
// ---------- File HelloWorldConnectionImpl.java
package book.jakartapro.jcahello;

import java.util.logging.Logger;
import book.jakartapro.jcahello.interfaces.
   HelloWorldConnection;

public class HelloWorldConnectionImpl
         implements HelloWorldConnection {
   private HelloWorldManagedConnection mc;
   private HelloWorldManagedConnectionFactory mcf;

   public HelloWorldConnectionImpl(
         HelloWorldManagedConnection mc,
         HelloWorldManagedConnectionFactory mcf) {
      this.mc = mc;
      this.mcf = mcf;
   }

   public String helloWorld() {
      return helloWorld(
         ((HelloWorldResourceAdapter) mcf.
         getResourceAdapter()).getName());
   }
```

```java
    public String helloWorld(String name) {
        return mc.helloWorld(name);
    }

    public void close() {
        mc.closeHandle(this);
    }
}

// ---------- File HelloWorldConnectionFactoryImpl.java
package book.jakartapro.jcahello;

import javax.naming.NamingException;
import javax.naming.Reference;

import book.jakartapro.jcahello.interfaces.
    HelloWorldConnection;
import book.jakartapro.jcahello.interfaces.
    HelloWorldConnectionFactory;
import jakarta.resource.ResourceException;
import jakarta.resource.spi.ConnectionManager;

public class HelloWorldConnectionFactoryImpl
        implements HelloWorldConnectionFactory {
    private Reference reference;
    private HelloWorldManagedConnectionFactory mcf;
    private ConnectionManager connectionManager;

    public HelloWorldConnectionFactoryImpl(
            HelloWorldManagedConnectionFactory mcf,
            ConnectionManager cxManager) {
        this.mcf = mcf;
        this.connectionManager = cxManager;
    }

    @Override
    public HelloWorldConnection getConnection()
            throws ResourceException {
        return (HelloWorldConnection) connectionManager.
```

```
        allocateConnection(mcf, null);
    }

    @Override
    public Reference getReference()
        throws NamingException {
      return reference;
    }

    @Override
    public void setReference(Reference reference) {
        this.reference = reference;
    }
}
```

The managed connection class and its factory class represent the physical connection to the EIS. The code reads as follows:

```
// ---------- File HelloWorldManagedConnection.java
package book.jakartapro.jcahello;

import java.io.PrintWriter;
import java.util.ArrayList;
import java.util.List;
import java.util.logging.Logger;

import jakarta.resource.NotSupportedException;
import jakarta.resource.ResourceException;
import jakarta.resource.spi.ConnectionEvent;
import jakarta.resource.spi.ConnectionEventListener;
import jakarta.resource.spi.ConnectionRequestInfo;
import jakarta.resource.spi.LocalTransaction;
import jakarta.resource.spi.ManagedConnection;
import jakarta.resource.spi.ManagedConnectionMetaData;

import javax.security.auth.Subject;
import javax.transaction.xa.XAResource;

import book.jakartapro.jcahello.interfaces.
```

```
        HelloWorldConnection;

public class HelloWorldManagedConnection
        implements ManagedConnection {
    private HelloWorldManagedConnectionFactory mcf;
    private List<ConnectionEventListener> listeners;
    private Object connection;
    private PrintWriter logWriter;

    public HelloWorldManagedConnection(
        HelloWorldManagedConnectionFactory mcf) {
        this.mcf = mcf;
        this.logWriter = null;
        this.listeners =
            new ArrayList<ConnectionEventListener>(1);
        this.connection = null;
    }

    /**
     * Creates a new connection handle for the underlying
     * physical connection represented by the
     * ManagedConnection instance.
     */
    public Object getConnection(
        Subject subject,
        ConnectionRequestInfo cxRequestInfo)
        throws ResourceException {
        connection = new HelloWorldConnectionImpl(this, mcf);
        return connection;
    }

    /**
     * Used by the container to change the association of
     * an application-level connection handle with a
     * ManagedConneciton instance.
     */
    public void associateConnection(Object connection)
```

357

```java
        throws ResourceException {
      this.connection = connection;
  }

  /**
   * Application server calls this method to force any
   * cleanup on the ManagedConnection instance.
   */
  public void cleanup() throws ResourceException {
  }

  /**
   * Destroys the physical connection to the underlying
   * resource manager.
   */
  public void destroy() throws ResourceException {
      this.connection = null;
  }

  public void addConnectionEventListener(
        ConnectionEventListener listener) {
      listeners.add(listener);
  }

  public void removeConnectionEventListener(
        ConnectionEventListener listener) {
      listeners.remove(listener);
  }

  public PrintWriter getLogWriter()
        throws ResourceException {
      return logWriter;
  }

  public void setLogWriter(PrintWriter out)
        throws ResourceException {
      this.logWriter = out;
  }
```

```java
    public LocalTransaction getLocalTransaction()
          throws ResourceException {
        throw new NotSupportedException(
            "LocalTransaction not supported");
    }

    public XAResource getXAResource()
          throws ResourceException {
        throw new NotSupportedException(
            "GetXAResource not supported");
    }

    public ManagedConnectionMetaData getMetaData()
          throws ResourceException {
        return new HelloWorldManagedConnectionMetaData();
    }

    String helloWorld(String name) {
        return "Hello World, " + name + " !";
    }

    void closeHandle(HelloWorldConnection handle) {
        ConnectionEvent event =
            new ConnectionEvent(this,
                ConnectionEvent.CONNECTION_CLOSED);
        event.setConnectionHandle(handle);

        for (ConnectionEventListener cel : listeners) {
            cel.connectionClosed(event);
        }
    }
}

// ---------- File HelloWorldManagedConnectionFactory.java
package book.jakartapro.jcahello;

import java.util.Objects;
import java.io.PrintWriter;
import java.util.Iterator;
```

```java
import java.util.Set;
import java.util.logging.Logger;

import jakarta.resource.ResourceException;
import jakarta.resource.spi.ConnectionDefinition;
import jakarta.resource.spi.ConnectionManager;
import jakarta.resource.spi.ConnectionRequestInfo;
import jakarta.resource.spi.ManagedConnection;
import jakarta.resource.spi.ManagedConnectionFactory;
import jakarta.resource.spi.ResourceAdapter;
import jakarta.resource.spi.ResourceAdapterAssociation;

import javax.security.auth.Subject;

import book.jakartapro.jcahello.interfaces.
    HelloWorldConnection;
import book.jakartapro.jcahello.interfaces.
    HelloWorldConnectionFactory;

@ConnectionDefinition(
  connectionFactory =
    HelloWorldConnectionFactory.class,
  connectionFactoryImpl =
    HelloWorldConnectionFactoryImpl.class,
  connection =
     HelloWorldConnection.class,
  connectionImpl =
    HelloWorldConnectionImpl.class)
public class HelloWorldManagedConnectionFactory
        implements ManagedConnectionFactory,
                    ResourceAdapterAssociation {
    private ResourceAdapter ra;
    private PrintWriter logwriter;

    public HelloWorldManagedConnectionFactory() {
        this.ra = null;
        this.logwriter = null;
    }
```

```java
public Object createConnectionFactory()
    throws ResourceException {
    throw new ResourceException(
        "This resource adapter doesn't " +
        "support non-managed environments");
}

public Object createConnectionFactory(
    ConnectionManager cxManager)
    throws ResourceException {
    return new HelloWorldConnectionFactoryImpl(this,
        cxManager);
}

public ManagedConnection createManagedConnection(
    Subject subject,
    ConnectionRequestInfo cxRequestInfo)
    throws ResourceException {
    return new HelloWorldManagedConnection(this);
}

public ManagedConnection matchManagedConnections(
    Set connectionSet, Subject subject,
    ConnectionRequestInfo cxRequestInfo)
    throws ResourceException {
    ManagedConnection result = null;
    Iterator it = connectionSet.iterator();
    while (result == null && it.hasNext()) {
        ManagedConnection mc =
            (ManagedConnection) it.next();
        if (mc instanceof
            HelloWorldManagedConnection) {
            HelloWorldManagedConnection hwmc =
                (HelloWorldManagedConnection) mc;
            result = hwmc;
```

```
            }
        }
        return result;
    }

    public PrintWriter getLogWriter()
            throws ResourceException {
        return logwriter;
    }

    public void setLogWriter(PrintWriter out)
            throws ResourceException {
        logwriter = out;
    }

    public ResourceAdapter getResourceAdapter() {
        return ra;
    }

    public void setResourceAdapter(ResourceAdapter ra) {
        this.ra = ra;
    }

    @Override
    public int hashCode() {
        return Objects.hash(17);
    }

    @Override
    public boolean equals(Object other) {
        if (other == null) return false;
        if (other == this) return true;
        if (!(other instanceof
                HelloWorldManagedConnectionFactory))
            return false;
        return true;
    }
}
```

The metadata object describes the resource adapter from a technical point of view. It reads as follows:

```java
// ---------- File HelloWorldManagedConnectionMetaData.java
package book.jakartapro.jcahello;

import jakarta.resource.ResourceException;
import jakarta.resource.spi.ManagedConnectionMetaData;

public class HelloWorldManagedConnectionMetaData
        implements ManagedConnectionMetaData {
    @Override
    public String getEISProductName()
          throws ResourceException {
        return "HelloWorld Resource Adapter";
    }

    @Override
    public String getEISProductVersion()
          throws ResourceException {
        return "1.0";
    }

    @Override
    public int getMaxConnections()
          throws ResourceException {
        return 0;
    }

    @Override
    public String getUserName()
          throws ResourceException {
        return null;
    }
}
```

Finally, the example's resource adapter class reads as follows:

```java
// ---------- File HelloWorldResourceAdapter.java
```

```java
package book.jakartapro.jcahello;

import java.util.logging.Logger;
import java.util.Objects;

import jakarta.resource.ResourceException;
import jakarta.resource.spi.ActivationSpec;
import jakarta.resource.spi.BootstrapContext;
import jakarta.resource.spi.ConfigProperty;
import jakarta.resource.spi.Connector;
import jakarta.resource.spi.ResourceAdapter;
import jakarta.resource.spi.
    ResourceAdapterInternalException;
import jakarta.resource.spi.TransactionSupport;
import jakarta.resource.spi.endpoint.
    MessageEndpointFactory;

import javax.transaction.xa.XAResource;

@Connector(
    reauthenticationSupport = false,
    transactionSupport =
    TransactionSupport.TransactionSupportLevel.
        NoTransaction)
public class HelloWorldResourceAdapter
        implements ResourceAdapter {
    /** Name property */
    @ConfigProperty(defaultValue = "Some Name",
        supportsDynamicUpdates = true)
    private String name;

    public void setName(String name) {
        this.name = name;
    }

    public String getName() {
        return name;
    }
```

```java
/**
 * This is called during the activation of a message
 * endpoint.
 */
public void endpointActivation(
      MessageEndpointFactory endpointFactory,
      ActivationSpec spec)
      throws ResourceException {
}

/**
 * This is called when a message endpoint is
 * deactivated.
 */
public void endpointDeactivation(
      MessageEndpointFactory endpointFactory,
      ActivationSpec spec) {
}

/**
 * This is called when a resource adapter instance
 * is bootstrapped.
 */
public void start(BootstrapContext ctx)
      throws ResourceAdapterInternalException {
}

/**
 * This is called when a resource adapter instance
 * is undeployed or during application server shutdown.
 */
public void stop() {
}

/**
 * This method is called by the application server
 * during crash recovery.
```

```java
    */
    public XAResource[] getXAResources(
        ActivationSpec[] specs)
        throws ResourceException {
        return null;
    }

    @Override
    public int hashCode() {
        return Objects.hash(name);
    }

    @Override
    public boolean equals(Object other) {
        return Objects.equal(this, other);
        if (other == null) return false;
        if (other == this) return true;
        if (!(other instanceof
            HelloWorldResourceAdapter))
            return false;
        HelloWorldResourceAdapter obj =
            (HelloWorldResourceAdapter) other;
        return Objects.equals(name, other.getName);
    }
}
```

Packaging and Deploying a Resource Adapter

A resource adapter is packaged like a JAR file, but contrary to a standard JAR file, it must use a filename ending in .rar. The structure of a simple RAR archive is as follows:

```
some/package/ClassA.class
some/other/package/ClassB.class
... more classes ...
META-INF/MANIFEST.MF
```

The MANIFEST.MF file is a simple text file with these contents:

```
Manifest-Version: 1.0
```

For the HelloJca example resource adapter, the archive file structure is as follows:

```
META-INF/
  MANIFEST.MF
book/
  jakartapro/
    jcahello/
      HelloWorldManagedConnection.class
      HelloWorldManagedConnectionFactory.class
      HelloWorldResourceAdapter.class
      HelloWorldConnectionFactoryImpl.class
      HelloWorldConnectionImpl.class
      HelloWorldManagedConnectionMetaData.class
      interfaces/
        HelloWorldConnectionFactory.class
        HelloWorldConnection.class
```

Because clients need the interfaces to address the resource adapter, you need to also set up a build workflow that packages the interfaces in a standard Java library JAR file. For the HelloJca example, this could be a file called jca-interfaces.jar containing:

```
META-INF/
  MANIFEST.MF
book/
  jakartapro/
    jcahello/
      interfaces/
        HelloWorldConnectionFactory.class
        HelloWorldConnection.class
```

For more complex resource adapters that need additional libraries, you could for example add those inside a lib/ folder in the RAR file:

```
some/package/ClassA.class
some/other/package/ClassB.class
... more classes ...
```

```
META-INF/MANIFEST.MF
lib/someLibrary1.jar
lib/someLibrary2.jar
... more libraries ...
```

The additional libraries then need to be mentioned in the MANIFEST.MF file:

```
Manifest-Version: 1.0
Class-Path: lib/someLibrary1.jar lib/someLibrary2.jar
```

(This should be on one line, with a single space as a separator.)

Caution   Make sure the MANIFEST.MF file ends with a line terminator (a new line or carriage return).

If you use Gradle as your build tool, changing the archive file ending to .rar and adding a step to produce a jca-interfaces.jar file is easy. You simply add the following lines to your build.gradle file:

```
plugins {
    id 'java'
}
...
archivesBaseName = 'HelloJca'
jar { archiveExtension = 'rar' }

repositories {
    ...
}

dependencies {
    ...
}

// Build a JAR with just the JCA interfaces, for clients
task('JcaInterfaces', type: Jar, dependsOn: 'classes') {
    //describe jar contents here as a CopySpec
    archiveFileName = "jca-interfaces.jar"
```

```
    destinationDirectory = file("$buildDir/libs")
    from("$buildDir/classes/java/main") {
        include "**/interfaces/**/*.*"
    }
}
// make sure it is part of the assembly path
jar.dependsOn JcaInterfaces
```

If you now execute the jar task, you will find both the RAR file and the jca-interfaces.jar file in the build/libs folder.

Deployment Descriptors

In older versions of JCA, it was necessary to provide a ra.xml deployment descriptor file inside the META-INF folder. Because of the annotations you can add to the resource adapter file (the HelloWorldResourceAdapter class in this example) and to the Managed Connection Factory (the HelloWorldManagedConnectionFactory class in the example), this file no longer is needed. However, depending on which Jakarta EE server you are using, a server-specific deployment descriptor has to be added to the META-INF folder. It usually has the name [SERVER]-ra.xml, where [SERVER] identifies the server product. For example, for Weblogic, this file must be named weblogic-ra.xml. For details about the naming and the contents of this file, consult your server documentation.

For GlassFish, this server-specific deployment descriptor is not needed—everything can be configured using the web administrator or the asadmin console tool. See the section, "Defining a Resource Adapter on the Server," later in this chapter.

Resource Adapter Deployment

To deploy a RAR file, you usually follow the same procedure as for a WAR or EAR file. Consult your server documentation for details.

Defining a Resource Adapter on the Server

Deploying a resource adapter on the server does not automatically mean you can instantaneously refer to it from a client. In the end, you want to be able to inject a resource adapter using the @Resource annotation, but that requires a lot of configuration work. Even worse, the configuration procedure is not standardized, so you have to consult the server documentation in order to learn how to do that.

For GlassFish, for example, you must create a thread pool, an adapter configuration, a connection pool, and a connector resource. You can use the web administrator at http://localhost:4848 to that aim, but a shell script using the asadmin tool gives you the same power. For the HelloJca example, such a script could read as follows:

```
# Create a new thread pool named "HelloJcaThreadPool"
bin/asadmin create-threadpool HelloJcaThreadPool

# Create a resource adapter configuration. The property
# "name" corresponds to a field in the resource
# adapter class (the one that inherits from
# jakarta.resource.spi.ResourceAdapter)
# The thread pool id must match the thread pool name
# used above. "HelloJca" will be the name of the resource
# adapter.
bin/asadmin create-resource-adapter-config \
   --property name="Gandalf" \
   --threadpoolid HelloJcaThreadPool \
   HelloJca

# Create a connection pool. The raname corresponds to the
# name of the resource adapter used above. The
# connectiondefinition corresponds to the connection
# factory class (the one that inherits from
# jakarta.resource.spi.ManagedConnectionFactory).
# "eis/HelloJcaConnectionPool" will be the JNDI name
# of the connectio pool.
bin/asadmin create-connector-connection-pool \
   --raname HelloJca \
   --connectiondefinition \
```

```
  book.jakartapro.jcahello.interfaces.
  HelloWorldConnectionFactory \
 eis/HelloJcaConnectionPool

# Create a connector resource. The poolname corresponds
# to the connection pool's JNDI name used above. The
# "eis/HelloWorld" will be the name to be used in
# an injectable field annotated with @Resource:
#   @Resource(name = "eis/HelloWorld")
#   private HelloWorldConnectionFactory connectionFactory;
# The field's type must of course be adapted if you use
# a different connection factory class.
bin/asadmin create-connector-resource \
  --poolname eis/HelloJcaConnectionPool \
  eis/HelloWorld
```

(There should be no line break and no spaces after `interfaces`.) For details and more options, see the GlassFish server documentation.

Resource Adapter Clients

Once the resource adapter is deployed and correctly configured, you then have to inject the Connection Factory via a `@Resource` annotation (the `jakarta.annotation` package) so the client can access the resource adapter's functionalities. For a REST controller, for example, the code would read as follows:

```
@Path("/")
public class HelloJca {
    @Resource(name = "eis/HelloWorld")
    private HelloWorldConnectionFactory connectionFactory;

    // more fields...

    @GET
    public Response fetch() {
        try {
            HelloWorldConnection conn = connectionFactory.
```

```
                getConnection();
        String s = conn.helloWorld();
        conn.close();
        return Response.status(200).
                entity("done: " + s).build();
    } catch (ResourceException e) {
        return Response.status(500).entity("error").
                build();
    }
  }
  ...
}
```

The name parameter of the annotation must match the JNDI name of the resource adapter.

CHAPTER 29

Caching

Accessing data from a large database or using complex `select` clauses can be time consuming tasks. In addition, it quite often happens that your application uses the same database access statement several times, which might make you wonder if there is a way to shortcut such database queries. Might that shortcut help the application react faster to user queries and avoid too much load on the database server? This is exactly what *caching* is for. It sits between the client code that runs database queries and the database access itself, and it answers database queries on behalf of the database when the same query has been run before.•

This chapter covers using Ehcache for the Hibernate persistence provider.

Installing Ehcache

In order to install Ehcache, add the following dependency to your build file:

```
// For Gradle:
...
dependencies {
    ...
    // Assuming you use the following Hibernate version
    implementation 'org.hibernate:' +
        'hibernate-core-jakarta:5.6.15.Final'
    // ... you must add:
    implementation 'org.hibernate:' +
        'hibernate-ehcache:5.6.15.Final'
}

<!-- For Maven: -->
...
```

© Peter Späth 2023
P. Späth, *Pro Jakarta EE 10*, https://doi.org/10.1007/978-1-4842-8214-4_29

```
<dependencies>
    ...
    <dependency>
      <groupId>org.hibernate</groupId>
      <artifactId>hibernate-ehcache</artifactId>
      <version>5.6.15.Final</version>
    </dependency>
</dependencies>
```

Configuring Hibernate for Ehcache

Before caching is actually enabled, you need to tell Hibernate that you want to use Ehcache. The Hibernate configuration for Jakarta EE happens inside the persistence. xml file, and this is also where you announce Ehcache to Hibernate:

```
<?xml version="1.0" encoding="UTF-8"?>
  <persistence ...>
    <persistence-unit ...>
      <!-- Make sure Hibernate gets used -->
      <provider>
          org.hibernate.jpa.HibernatePersistenceProvider
      </provider>
      ...
      <properties>
        ...
        <property
          name="hibernate.cache.use_second_level_cache"
          value="true"/>
        <property
          name="hibernate.cache.use_query_cache"
          value="true"/>
        <property
          name="hibernate.cache.region.factory_class"
          value="org.hibernate.cache.ehcache.
                  SingletonEhCacheRegionFactory"/>
        <property
```

```
            name="hibernate.cache.ehcache.
                missing_cache_strategy"
            value="create"/>
        <property name="hibernate.generate_statistics"
            value="true"/>
        <property name="hibernate.jmx.enabled"
            value="true"/>
        <property name="hibernate.jmx.usePlatformServer"
            value="true"/>
    </properties>
  </persistence-unit>
</persistence>
```

(Be sure to remove the newlines and the spaces inside the "...".) A few notes on the properties seem appropriate; for details, consult the documentation at https://www.ehcache.org.

- hibernate.cache.use_second_level_cache

  Set this to true if you want the second level cache to be enabled. The second level cache stores entities by ID in a session agnostic way, which means it works across all sessions. The default is false.

- hibernate.cache.use_query_cache

  Set this to true if you want the query cache to be enabled. The query cache is used to store the results of explicit HQL queries in the code, but only if you add .setCacheable(true) while building a Criteria object. The default is false.

- hibernate.cache.region.factory_class

  Defines the region factory class used for caching. The other possible value is org.hibernate.cache.ehcache.EhCacheRegionFactory.

- hibernate.cache.ehcache.missing_cache_strategy

  If you don't explicitly define regions in a class' Cache annotation (the region parameter), a create value for this property tells Hibernate to create a region based on the class name. Otherwise, you'll get a warning in the logs.

- hibernate.generate_statistics

    Set this to true if you want Hibernate to generate statistics data, including caching usage. The default is false.

- hibernate.jmx.enabled

    Set this to true if you want Hibernate to announce statistical figures via JMX. The default is false.

- hibernate.jmx.usePlatformServer

    Set this to true if you want Hibernate to use the platform MBean server for JMX. The default is false.

Furthermore, you need a META-INF/ehcache.xml file, which is where the caching parameters are defined. An example is as follows:

```xml
<?xml version="1.0" encoding="UTF-8"?>
<ehcache
  xmlns:xsi="http://www.w3.org/2001/XMLSchema-instance"
  xsi:noNamespaceSchemaLocation="ehcache.xsd"
  updateCheck="true"
  monitoring="autodetect"
  dynamicConfig="true">
<diskStore path="java.io.tmpdir/ehcache" />
<defaultCache
    maxElementsInMemory="1000"
    eternal="false"
    timeToIdleSeconds="3"
    timeToLiveSeconds="120"
    overflowToDisk="true"
    statistics="true" />
<cache name="Region1"
    maxElementsInMemory="100"
    eternal="false"
    timeToIdleSeconds="3"
    timeToLiveSeconds="120"
```

```
  overflowToDisk="false"
  statistics="true" />
</ehcache>
```

The Ehcache documentation readily tells you all about the definitions and parameters you can set in this file.

Defining Cached Objects in the Code

Hibernate in the end sends SQL statements, but because it is an ORM (object relational mapper), the code refers to cached Java objects. However, Hibernate does not cache all objects by default. Instead, you have to add the org.hibernate.annotations.Cache annotation to the JPA entity class:

```
package ...;

import org.hibernate.annotations.Cache;
import org.hibernate.annotations.CacheConcurrencyStrategy;
import jakarta.persistence.Entity;
import jakarta.persistence.Table;

@Entity
@Table(name = "person")
@Cache(usage=CacheConcurrencyStrategy.READ_ONLY,
       region="Region1")
public class Person {
    .... some JPA class
}
```

The region parameter of the @Cache annotation corresponds to a region defined in the ehcache.xml file. The usage parameter specifies the concurrency strategy. You can use one of the following for its value:

- CacheConcurrencyStrategy.NONE

    No concurrency strategy. Hibernate doesn't care which application updates or reads data. If there are cached objects, they are used irrespective of whether there are other parties accessing the same table.

377

- CacheConcurrencyStrategy.READ_ONLY

  Use this to indicate that your application only reads from and never writes to this table.

- CacheConcurrencyStrategy.READ_WRITE

  Use this to indicate that your application reads from and writes to this table.

- CacheConcurrencyStrategy.NONSTRICT_READ_WRITE

  Use this to indicate that your application reads from this table and only once in a while writes to it as well.

- CacheConcurrencyStrategy.TRANSACTIONAL

  Use this to indicate that the caching manager allows for transactions. This book doesn't explore transactional caching; for details, see the "Transactions in Ehcache" section of the Ehcache documentation.

Monitoring Caching Activities

Because you enabled statistical data gathering and JMX access in the former sections, you can use a JMX browser to investigate the cache performance. You can use the VisualVM program at https://visualvm.github.io/ for this purpose. After you install it and add the MBeans plugin, navigate to Hibernate's statistics page, as shown in Figure 29-1.

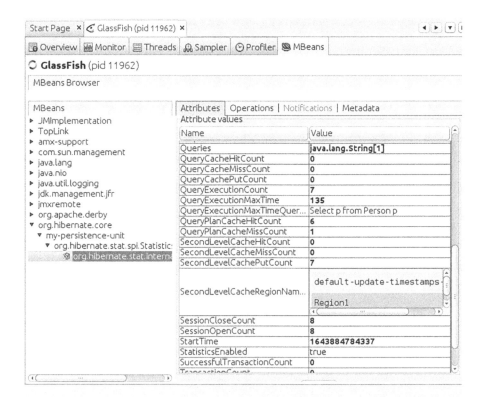

Figure 29-1. *VisualVM JMX access*

## CHAPTER 30

NoSQL

SQL databases do a good job at mapping real-world relations to data storages. However, developing a big relationship structure is not an easy task, because later in a project or during the production stage, changing relations and adding or moving table columns are time-consuming tasks. You have to check whether database entry relations are still valid and correctly correlate to database access code, and check whether all the clients can cope with an updated database structure. For these reasons, a considerable amount of effort is spent at the beginning of a project to create a bullet-proof database relation model.

In order to simplify structural data-model changes, you have to ease or even abandon strict database table relations. Such database management systems have recently gained a considerable amount of attention when used in modern usage scenarios, where database structures change often. They provide a usable API from Java clients. They are commonly referred to as "NoSQL" databases. This is an abbreviation of "Not-Only-SQL," because often the traditional way of accessing data in a SQL-like manner still is possible, while schema-less access to database entries is the preferred way of addressing table entries.

There is no finalized Jakarta EE 10 standard describing access to NoSQL databases (see `https://jakarta.ee/specifications/nosql/` for the ongoing work). Nevertheless, because of the importance that NoSQL databases have today, this chapter briefly describes the two NoSQL DBMSs—CouchDB and MongoDB.

Using CouchDB from Jakarta EE Applications

CouchDB provides a REST API that you can address from Java using a `HttpClient` from the `java.net.http` package. Assuming you have a CouchDB running locally on port 5984 and an admin user `admin` with password `PASSWD`, you first define a couple of helper functions:

© Peter Späth 2023
P. Späth, *Pro Jakarta EE 10*, https://doi.org/10.1007/978-1-4842-8214-4_30

```java
import java.net.Authenticator;
import java.net.PasswordAuthentication;
import java.net.URI;
import java.net.http.HttpClient;
import java.net.http.HttpRequest;
import java.net.http.HttpResponse;
import java.net.http.HttpResponse.BodyHandlers;
import java.nio.charset.StandardCharsets;
import java.time.Duration;
import java.util.Base64;
import java.util.function.BiFunction;
import java.util.function.Function;
...

final String SERVER = "127.0.0.1:5984";
final String USER = "admin";
final String PASSWD = "PASSWD";

var auth = new Authenticator() {
    protected PasswordAuthentication
    getPasswordAuthentication() {
        return new PasswordAuthentication(USER,
                PASSWD.toCharArray());
    }
};
var encodedAuth = Base64.getEncoder()
    .encodeToString((USER + ":" + PASSWD)
    .getBytes(StandardCharsets.UTF_8));

Function<String,HttpRequest> get = (url) -> {
    return HttpRequest.newBuilder()
        .uri(URI.create("http://" + SERVER + "/" + url))
        .timeout(Duration.ofSeconds(10))
        .header("Authorization", "Basic " + encodedAuth)
        .GET().build();
};
```

```java
Function<String,HttpRequest> delete = (url) -> {
    return HttpRequest.newBuilder()
        .uri(URI.create("http://" + SERVER  + "/" + url))
        .timeout(Duration.ofSeconds(10))
        .header("Authorization", "Basic " + encodedAuth)
        .DELETE().build();
};

BiFunction<String,String,HttpRequest> post =
(url,body) -> {
    return HttpRequest.newBuilder()
        .uri(URI.create("http://" + SERVER + "/" + url))
        .timeout(Duration.ofSeconds(10))
        .header("Authorization", "Basic " + encodedAuth)
        .POST(HttpRequest.BodyPublishers.ofString(body))
        .build();
};

BiFunction<String,String,HttpRequest> put = (url,body) -> {
    return HttpRequest.newBuilder()
        .uri(URI.create("http://" + SERVER + "/" + url))
        .timeout(Duration.ofSeconds(10))
        .header("Authorization", "Basic " + encodedAuth)
        .PUT(HttpRequest.BodyPublishers.ofString(body))
        .build();
};

HttpClient client = HttpClient.newBuilder()
        .version(HttpClient.Version.HTTP_1_1)
        .followRedirects(HttpClient.Redirect.NORMAL)
        .connectTimeout(Duration.ofSeconds(10))
        .authenticator(auth)
        .build();

Function<HttpRequest,HttpResponse<String>> invoke =
(req) -> {
    try {
```

```
        return client.send(req, BodyHandlers.ofString());
    } catch (Exception e) {
        e.printStackTrace(System.err);
        return null;
    }
};
```

For simplicity, this example uses lambda expressions, but the helper functions can readily be translated to class methods.

It is now easy to send requests to the CouchDB instance. Search statements are deliberately not used in the following code snippet, since for CouchDB, searchable indexes need to be created and/or search plugins need to be installed, which is out of scope of this brief example.

```
// ----- CouchDB info
var request1 = get.apply("");
var response1 = invoke.apply(request1);
System.out.println(response1.statusCode() +"\n"
    + response1.body());

// ----- create database
var request2 = put.apply("mydb","");
var response2 = invoke.apply(request2);
System.out.println(response2.statusCode() + "\n"
    + response2.body());

// ----- create entry
var request3 = put.apply("mydb/001",
    "{ \"name\":\"John\", \"age\":42 }");
var response3 = invoke.apply(request3);
System.out.println(response3.statusCode() + "\n"
    + response3.body());

// ----- delete database
var request99 = delete.apply("mydb");
var response99 = invoke.apply(request99);
System.out.println(response99.statusCode() + "\n"
    + response99.body());
```

Using MongoDB from Jakarta EE Applications

MongoDB has a dedicated Java driver. Add the following dependencies to your project (this is for Gradle):

```
...
dependencies {
    ...
    implementation 'org.mongodb:' +
        'mongodb-driver-sync:4.5.1'
    implementation 'org.slf4j:' +
        'slf4j-simple:1.8.0-beta4'
    implementation 'org.slf4j:' +
        'slf4j-api:1.8.0-beta4'
}
```

For Maven dependencies, write this inside the pom.xml file:

```
...
<dependencies>
    ...
    <dependency>
      <groupId>org.mongodb</groupId>
      <artifactId>mongodb-driver-sync</artifactId>
      <version>4.5.1</version>
    </dependency>
    <dependency>
      <groupId>org.slf4j</groupId>
      <artifactId>slf4j-api</artifactId>
      <version>1.8.0-beta4</version>
    </dependency>
    <dependency>
      <groupId>org.slf4j</groupId>
      <artifactId>slf4j-simple</artifactId>
      <version>1.8.0-beta4</version>
    </dependency>
</dependencies>
```

In the code, you can then use the classes from the driver API. Note that MongoDB automatically creates databases and collections (tables in MongoDB are called "collections") upon first usage if they don't exist. So you don't have to explicitly create them.

```java
...
import org.bson.Document;
import org.bson.types.ObjectId;

import com.mongodb.client.MongoClient;
import com.mongodb.client.MongoClients;
import com.mongodb.client.MongoCollection;
import com.mongodb.client.MongoDatabase;

public class MongoMain {
    public static void main(String[] args) {
        MongoClient client = MongoClients.create();
        // Accessing a database:
        MongoDatabase database = client.
            getDatabase("my_database");
        // Accessing a collection (table):
        MongoCollection<Document> persons =
            database.getCollection("persons");
        // Create a document (an entry):
        Document p = new Document("name", "John")
            .append("cars", new Document("Tesla", 2019));
        // Save it:
        ObjectId id = persons.insertOne(p)
            .getInsertedId().asObjectId().getValue();
        // Search for a document (an entry):
        Document john = persons.find(
            new Document("name", "John")).first();
        System.out.println(john.toJson());
    }
}
```

This is for a MongoDB instance running on the localhost and using the standard port 27017. For different addresses or ports, look at `MongoClients'` other `create( ... )` methods. For more details, see `https://docs.mongodb.com/drivers/java/sync/current/` and `https://mongodb.github.io/mongo-java-driver/`.

PART VI

Security Enhancements

CHAPTER 31

Secured JMX

JMX (Java Management Extensions) is a very powerful and largely underestimated way to monitor and control a Jakarta EE application server or the applications running inside it. I do not provide an introduction to JMX here, because you can find lots of stuff about it on the Internet, including Oracle's documentation at `https://docs.oracle.com/javase/tutorial/jmx/overview/index.html`.

What I do instead in this chapter is explain how to secure remote JMX connections using SSL, because this might come up in a corporate environment.

Using SSL for Remote JMX Connections

In order to use SSL for JMX connections, you need a key and a trust store on both sides of the server (Jakarta EE server) and the client (a remote tool for investigating or monitoring JMX). The client key's public part is entered into the server's trust store and vice versa, so the client and server authenticate themselves at the communication partner.

The following sections describe how to generate the keys and trust stores, and how to configure client and server accordingly.

Generating the SSL Keys

For key and certificate store management, you must use the `keytool` CLI utility, which is part of every JDK installation. The following bash script does all the work:

```
#!/bin/bash
PASSWORD=dhdgdfd534
DNAME="cn=JMX, ou=Java, o=Glassfish"

# ############################################################
```

© Peter Späth 2023
P. Späth, *Pro Jakarta EE 10*, https://doi.org/10.1007/978-1-4842-8214-4_31

```
# JMX server
# ############################################################
DAYS=3650
SERVERCACERTS="/path/to/server/conf/cacerts.jks"

rm -rf jmxserver.keystore jmxserver.truststore

#Make a key
keytool -genkey -alias serverjmx -keyalg RSA \
    -validity ${DAYS} \
    -keystore jmxserver.keystore -storepass ${PASSWORD} \
    -keypass ${PASSWORD} -dname "${DNAME}"

#Make our own copy of the Truststore
cp ${SERVERCACERTS} jmxserver.truststore
#Convert to new format (Glassfish)
keytool -importkeystore \
    -srckeystore jmxserver.truststore \
    -srcstorepass changeit \
    -destkeystore jmxserver.truststore \
    -deststorepass changeit \
    -deststoretype pkcs12
rm -rf jmxserver.truststore.old
keytool -storepasswd -keystore jmxserver.truststore \
    -storepass changeit -new ${PASSWORD}

#Make a new key inside the truststore
keytool -genkey -alias serverjmx -keyalg RSA \
    -validity ${DAYS} -keystore jmxserver.truststore \
    -storepass ${PASSWORD} \
    -dname "${DNAME}"

# ############################################################
# JMX client
# ############################################################
DAYS=3650

# Make an empty truststore
```

```
keytool -genkeypair -alias boguscert \
    -storepass ${PASSWORD} -keystore emptyStore.keystore \
    -dname "CN=Developer, OU=Department, O=Company, \
    L=City, ST=State, C=CA"
keytool -delete -alias boguscert -storepass ${PASSWORD} \
    -keystore emptyStore.keystore
mv emptyStore.keystore "${CLIENTCACERTS}"

rm -rf jmxclient.keystore jmxclient.truststore

#Make a key
keytool -genkey -alias clientjmx -keyalg RSA \
    -validity ${DAYS} -keystore jmxclient.keystore \
    -storepass ${PASSWORD} \
    -keypass ${PASSWORD} \
    -dname "${DNAME}"

#Make a new key inside the truststore
keytool -genkey -alias jmxclient -keyalg RSA \
    -validity ${DAYS} -keystore "${CLIENTCACERTS}" \
    -storepass ${PASSWORD} -dname "${DNAME}"

###############################################################
#Export the public certificates from the keystores:
###############################################################
rm -rf jmxserver.cer jmxclient.cer
keytool -export -alias serverjmx \
    -keystore jmxserver.keystore -file jmxserver.cer \
    -storepass ${PASSWORD}
keytool -export -alias clientjmx \
    -keystore jmxclient.keystore -file jmxclient.cer \
    -storepass ${PASSWORD}

###############################################################
#Finally, import the certificates into each other's
#truststore. This allows the server to trust the client,
#and vice versa:
```

```
###########################################################
keytool -import -alias jmxclient -file jmxclient.cer \
    -keystore jmxserver.truststore \
    -storepass ${PASSWORD} -noprompt

keytool -import -alias jmxserver -file jmxserver.cer \
    -keystore jmxclient.truststore \
    -storepass ${PASSWORD} -noprompt

###########################################################
#Get rid of any remaining CER certificate files
###########################################################
rm -f jmxserver.cer jmxclient.cer
```

At the top of the script, you must define a password for all key and trust stores. You should use your own, of course. DNAME is a distinguished name to be used for all keys—you can adapt it to your needs, if you like. Then you build a server key and trust store, and a client key and trust store. After extracting the public part of each of the keys from the keystores, you register the client's public certificate with the server's trust store, and the server's public certificate with the client's trust store. The whole process leaves you with the following files:

- `jmxserver.keystore`

  Contains the key used by the server to identify itself. Also, its public part is registered with the client's trust store.

- `jmxserver.truststore`

  Contains the public keys of all clients that want to authenticate to the server. This includes the JMX client's public key.

- `jmxclient.keystore`

  Contains the key used by the client to identify itself. Also, its public part gets registered with the server's trust store.

- `jmxclient.truststore`

  Contains the public keys of all servers that want to authenticate to the client. This is usually the server's public key.

Configuring the Server

There is no standard for configuring JMX on a Jakarta EE server, so you have to consult the server's manual for that purpose. Often, there is a startup configuration file that can be used to set the system properties. If this is the case for you, the basic steps for configuring the server for JMX are first to copy or move the jmxserver.keystore and jmxserver.truststore files to a suitable place for the server installation. Then, make sure the following system properties are set:

```
javax.net.ssl.keyStore=
    /path/to/keystores/jmxserver.keystore
javax.net.ssl.trustStore=
    /path/to/keystores/jmxserver.truststore
com.sun.management.jmxremote=
    true
com.sun.management.jmxremote.port=
    9010
com.sun.management.jmxremote.authenticate=
    false
com.sun.management.jmxremote.ssl=
    true
javax.net.ssl.keyStorePassword=
    dhdgdfd534
javax.net.ssl.trustStorePassword=
    dhdgdfd534
com.sun.management.jmxremote.ssl.need.client.auth=
    true
```

Change the passwords and paths according to your needs. After that, you usually have to restart the server.

Configuring the Client

If the JMX client you want to use to monitor or investigate the JMX figures is a Java program, the steps to access the SSL-secured JMX are quite similar to those for the server. Likewise, the configuration differs from client to client, so again you have to consult the documentation for details. Regardless of your client software, first move or copy the jmxclient.keystore and jmxclient.truststore files to a suitable place inside the client installation. For the configuration itself, in many cases there is a script or configuration file where the following properties can be set:

```
javax.net.ssl.keyStore=
    /path/to/keystores/jmxclient.keystore
javax.net.ssl.keyStorePassword=
    dhdgdfd534
javax.net.ssl.trustStore=
    /path/to/keystores/jmxclient.truststore
javax.net.ssl.trustStorePassword=
    dhdgdfd534
```

For VisualVM, for example, after you install the MBeans plugin, locate the etc/visualvm.conf file and add the following right after the -J-Xmx768m string.:

```
-J-Djavax.net.ssl.keyStore=
    /path/to/keystores/jmxclient.keystore
-J-Djavax.net.ssl.keyStorePassword=
    dhdgdfd534
-J-Djavax.net.ssl.trustStore=
    /path/to/keystores/jmxclient.truststore
-J-Djavax.net.ssl.trustStorePassword=
    dhdgdfd534
```

(Make sure there are no line breaks or spaces after the = signs.)

With the configuration set up, you need a JMX URL to connect to the server. This URL differs from Jakarta EE server to Jakarta EE server, but for GlassFish, and as a starting point for other servers, use this URL:

```
service:jmx:rmi:///jndi/rmi://localhost:9010/jmxrmi
```

where localhost possibly needs to be replaced with the JMX server network address, and 9010 corresponds to the port declared in the server configuration.

Disabling Random JMX Ports

In the server configuration, this example declares a JMX port as follows:

```
com.sun.management.jmxremote.port=9010
```

The whole story is more complicated, though. The truth is, the specified port establishes JMX connections, it does not actually communicate with MBeans. This workload port is usually referred-to as RMI-Port. Without further configuration, the RMI-Port is a randomly assigned number. This happens under the hood, so from a development or configuration point of view, there is no need to care about this random port. In a corporate environment, however, a randomly assigned port frequently causes troubles, because a firewall might block the RMI communication.

For this reason, there is another setting you can use to assign a static port for JMX-RMI communication. You can even use the same port to establish a connection and use it. The second line in

```
...
com.sun.management.jmxremote.port=9010
com.sun.management.jmxremote.rmi.port=9010
...
```

declares a system property to set a static JMX-RMI port.

CHAPTER 32

Java Web Tokens with Encryption

Chapter 14 covered JWTs (Java Web Tokens), which are used to enhance RESTful client-server communication by adding security-related claims. That same chapter presented a signing process to ensure that the token's content hadn't been tampered with. For that purpose, you added the JJWT library to the web application. Using JJWT allows you to write code like the following as a token-generation method:

```java
import io.jsonwebtoken.Jwts;
...
    Key key = keyGenerator.generateKey();
    String jwtToken = Jwts.builder()
        .setSubject(login)
        .setIssuer(uriInfo.getAbsolutePath().toString())
        .setIssuedAt(new Date())
        .setExpiration(toDate(
            LocalDateTime.now().plusMinutes(15L)))
        .signWith(key, SignatureAlgorithm.HS512)
        .compact();
...
```

However, if security requirements go a step further and encryption needs to be added, you must watch out for a library that implements the JWE (Java Web Tokens with Encryption) specification. JJWT provides signing, not encryption, but there are a couple of projects that target JWE. One of them is called *Jose4j*, and you can download it and get its documentation at `https://bitbucket.org/b_c/jose4j/wiki/Home`.

© Peter Späth 2023
P. Späth, *Pro Jakarta EE 10*, https://doi.org/10.1007/978-1-4842-8214-4_32

This chapter briefly outlines JWE and Jose4j. For more details, see the project's home page and the JWE specification at `https://datatracker.ietf.org/doc/rfc7516/ballot/`.

Note   JWT, JWS, and JWE are not part of the Jakarta EE specification. I added it because of the attention that communication via JWTs receives nowadays.

Installing Jose4j

In order to install Jose4j, add the following dependency to your Gradle (`build.gradle`) or Maven (`pom.xml`) build file:

```
// Gradle:
...
dependencies {
  ...
  implementation 'org.bitbucket.b_c:jose4j:0.7.10'
}
<!-- Maven: -->
<dependencies>
  ...
  <dependency>
    <groupId>org.bitbucket.b_c</groupId>
    <artifactId>jose4j</artifactId>
    <version>0.7.10</version>
  </dependency>
</dependencies>
```

Encrypting Claims

Claims represent statements about an entity, typically a person, that needs to be asserted. So, to issue a JWT or an encrypted JWT, you first need to collect the claims:

```
import org.jose4j.jwt.JwtClaims;
```

...

```
    JwtClaims claims = new JwtClaims();
    claims.setIssuer("Issuer");
    // <- token creator
    claims.setIssuedAtToNow();
    // <- token created now
    claims.setAudience("Audience");
    // <- intended token recipient
    claims.setExpirationTimeMinutesInTheFuture(10);
    // <- expiry time
```

...

For more claim-related options, consult the API documentation.

As a next step, you need a key for encrypting the JWT. Naturally, there are many options for creating keys. You can use symmetric or asymmetric encryption, and there are several suitable encryption algorithms at hand. This example uses a simple but nevertheless strong symmetric key in the form of an octet stream (a string cipher):

```
import org.jose4j.jwk.JsonWebKey;
...
  String jwkJson = "{" +
     "\"kty\":\"oct\"," +
     "\"k\":" +
        "\"ffdbdh4grh5tffnkkj5tvrfcddehjihhhfddeesdfff\"" +
  "}";
  JsonWebKey jwk = JsonWebKey.Factory.newJwk(jwkJson);
```

Again, for alternatives and more options, the Jose4j documentation will readily help you.

Next, create a JsonWebEncryption object, add the claims to it, set a few parameters, and register the key.

```
import org.jose4j.jwa.AlgorithmConstraints;
import org.jose4j.jwa.AlgorithmConstraints.ConstraintType;
import org.jose4j.jwe.
     ContentEncryptionAlgorithmIdentifiers;
import org.jose4j.jwe.JsonWebEncryption;
```

...

```java
JsonWebEncryption jwe = new JsonWebEncryption();

// The payload of the JWE token
jwe.setPayload(claims.toJson());

// The "alg" header indicates the key management mode
// for this JWE.
// For more options, see API documentation
jwe.setAlgorithmHeaderValue(
        KeyManagementAlgorithmIdentifiers.DIRECT);

// The "enc" header indicates the content encryption
// algorithm to be used.
// For more options, see API documentation
jwe.setEncryptionMethodHeaderParameter(
    ContentEncryptionAlgorithmIdentifiers.
    AES_128_CBC_HMAC_SHA_256);

// Set the key on the JWE.
jwe.setKey(jwk.getKey());
```

With all that set up, you are now ready to build the JWE token:

```java
// Produces the JWE compact serialization. Here the
// actual encryption is done.
// See the JWE specification for details
String theJweToken = jwe.getCompactSerialization();
```

The string result might look like this (all on one line):

eyJhbGciOiJkaXIiLCJlbmMiOiJBMTI4Q0JDLUhTMjU2In0..
EfAfbcwEl-K1hvz-J8sleg.ijrS45pJLeX_pNdfHyRGD54Iml
3Sr-kvYL4RDJJD_4MwlADQspSeUxJm8skSyKtJCK_R_Lyim6B
rxt_3LbUHPsAQRmEb_dIw1NitjBw2aVM.-OELtMngH-n9WLGf
kOlPeQ

A JWE token consists of several Base64-encoded parts, delimited by a stop sign "." (a period). If you Base64-decode each part, you get the following:

{"alg":"dir","enc":"A128CBC-HS256"}
<empty>
<some binary data>
<some binary data>
<some binary data>

You can see that, apart from information about the algorithm, only encrypted data is included with the JWE token.

Decrypting Claims

On the receiver side, you must decrypt the JWE message to inspect its contents. You first build the same key that was used for encryption. This is characteristic of the process, since you use symmetric encryption. Then you construct a JsonWebEncryption object, set a few parameters on it, and then register the JWE message and the key with it.

```
import org.jose4j.jwa.AlgorithmConstraints;
import org.jose4j.jwa.AlgorithmConstraints.ConstraintType;
import org.jose4j.jwe.
      ContentEncryptionAlgorithmIdentifiers;
import org.jose4j.jwe.JsonWebEncryption;
import org.jose4j.jwe.KeyManagementAlgorithmIdentifiers;
import org.jose4j.jwk.JsonWebKey;
import org.jose4j.jwt.JwtClaims;
import org.jose4j.lang.JoseException;
...

  String incommingJwe = ...;

  String jwkJson = "{" +
     "\"kty\":\"oct\"," +
     "\"k\":" +
       "\"ffdbdh4grh5tffnkkj5tvrfcddehjihhhfddeesdfff\"" +
  "}";
  JsonWebKey jwk = JsonWebKey.Factory.newJwk(jwkJson);

  JsonWebEncryption jwe = new JsonWebEncryption();
```

```
// Set the algorithm constraints.
// For explanation, see API documentation
AlgorithmConstraints algConstraints =
  new AlgorithmConstraints(ConstraintType.PERMIT,
  KeyManagementAlgorithmIdentifiers.DIRECT);
jwe.setAlgorithmConstraints(
    algConstraints);
AlgorithmConstraints encConstraints =
  new AlgorithmConstraints(ConstraintType.PERMIT,
  ContentEncryptionAlgorithmIdentifiers.
      AES_128_CBC_HMAC_SHA_256);
jwe.setContentEncryptionAlgorithmConstraints(
    encConstraints);

// Register the JWE token
jwe.setCompactSerialization(incommingJwe);

// Set the key
receiverJwe.setKey(jwk.getKey());
```

It is now easy to perform the decryption and investigate the token claims:

```
// Decrypt the message.
String plaintext = jwe.getPlaintextString();
System.out.println("plaintext: " + plaintext);
```

The result shows the claims you entered on the sender side:

```
plaintext:
{
  "iss":"Issuer",
  "iat":1645887062,
  "aud":"Audience",
  "exp":1645887662
}
```

Further Reading

This chapter showed only a brief example about JWE and the Jose4j library. There is a lot more to say about JWE and its usage scenarios. A good starting point for further research on that subject is the specification document at `https://datatracker.ietf.org/doc/rfc7516/ballot/`. For a more hands-on approach, the Jose4j site at `https://bitbucket.org/b_c/jose4j/wiki/Home` contains valuable information and presents many comprehensive examples covering many aspects of JWE.

## CHAPTER 33

Java Enterprise Security

With the Java Enterprise Security API (version 3.0 for Jakarta EE 10), you have a standardized feature for the implementation of authorization mechanisms. This chapter shows a few glimpses of this technology. For more details, see the specification at `https://download.oracle.com/otn-pub/jcp/java_ee_security-1_0-final-eval-spec/jsr375-spec-1.0-final.pdf`.

Form-Based Authentication

You learned that you have to enter authentication information in the `web.xml` file, for example:

```
<login-config>
  <auth-method>FORM</auth-method>
  <realm-name>file</realm-name>
  <form-login-config>
    <form-login-page>
      /login/login.xhtml
    </form-login-page>
    <form-error-page>
      /login/error.xhtml
    </form-error-page>
  </form-login-config>
</login-config>
```

However, independent of the authentication method, the question is how the container can get the credentials, i.e., which component it can ask for the login and password in order to check for correct user input. By using the Security API, you can directly hook into the authentication and user data storage retrieval process.

© Peter Späth 2023
P. Späth, *Pro Jakarta EE 10*, https://doi.org/10.1007/978-1-4842-8214-4_33

The Security API

The Security API, starting with version 1.0, extends and simplifies the Java
Authentication Service Provider Interface for Containers (JASPIC/JSR-196). The latter for
example allowed for a customizable authentication independent of the container.

The Security API, governed by JSR-375, is part of Jakarta EE 10, so there is no need for
additional installation steps.

The following sections shed some light on the new annotations, classes, and
interfaces defined by the standard. In addition to being able to readily implement
custom authentication procedures, it is also possible to support "Remember Me" and
"Login to Continue" functionalities.

Authentication Data: IdentityStore

```java
public interface IdentityStore {
  CredentialValidationResult
      validate(Credential credential);
  Set<String>
      getCallerGroups(CredentialValidationResult result);
  int priority();
  Set<ValidationType> validationTypes();
  enum ValidationType { VALIDATE, PROVIDE_GROUPS }
}
```

The IdentityStore interface defines the source of authentication data. By
implementing the validationTypes() method, you can specify the responsibilities
for the identity store. It is possible to let it handle login data, role assignment, or both.
Possible values of elements in the return set are described by the ValidationType
enumeration:

- VALIDATE

  The IdentityStore is used to validate login data.

- PROVIDE_GROUPS

  The IdentityStore allows for the assignment of user groups (roles, in
  a sense).

The IdentityStore interface's API documentation describes the other methods.

Registering IdentityStores happens automagically by placing them in the CLASSPATH. From a development point of view, you implement the IdentityStore interface and that is it. With JSR-375-governed authentication methods, all registered identity stores are queried, until one of them returns a positive validation. After that, the identity stores are queried again to interrogate group assignments.

Handling multiple identity stores is controlled by an implementation of the IdentityStoreHandler interface. Normally you don't have to provide your own implementation of this interface, but you *can*, if you like.

When you use an LDAP server or a SQL database as storage for identity data, you don't have to provide a fully-fledged IdentityStore implementation. Instead, you can use one of the two annotations provided by the JSR-375 standard:

```
@LdapIdentityStoreDefinition(
  url = "ldap://localhost:4321",
  bindDn = "ldap@thecompany",
  bindDnPassword = "password"
)

@DatabaseIdentityStoreDefinition(
  dataSourceLookup = "java:comp/env/ExampleDS",
  callerQuery = "SELECT password from USERS where name = ?"
)
```

Just add them as an annotation to a type seen by CDI, as in

```
@DatabaseIdentityStoreDefinition(
  dataSourceLookup = "java:comp/env/ExampleDS",
  callerQuery = "SELECT password from USERS where name = ?"
)
@ApplicationScoped
public class ApplConfig {
}
```

Authentication Methods: HttpAuthenticationMechanism

For the authentication procedure itself, implementing the HttpAuthenticationMechanism interface allows you to control HTTP data flow to ensure that only authenticated users can access the secured content.

```java
public interface HttpAuthenticationMechanism {

  AuthenticationStatus validateRequest(
    HttpServletRequest request,
    HttpServletResponse response,
    HttpMessageContext httpMessageContext)
  throws AuthenticationException;

  AuthenticationStatus secureResponse(
    HttpServletRequest request,
    HttpServletResponse response,
    HttpMessageContext httpMessageContext)
  throws AuthenticationException;

  void cleanSubject(
    HttpServletRequest request,
    HttpServletResponse response,
    HttpMessageContext httpMessageContext)
  throws AuthenticationException;
}
```

During the authentication process, the validateRequest() method is invoked. This is where the login information (username/password) is validated, usually with the help of an injected IdentityStoreHandler:

```java
public class MyAuth
implements HttpAuthenticationMechanism {
  @Inject IdentityStoreHandler ish;

  AuthenticationStatus validateRequest(
    HttpServletRequest request,
```

```
  HttpServletResponse response,
  HttpMessageContext httpMessageContext)
throws AuthenticationException {
    // use ish to validate the request
    ...
    CredentialValidationResult vr = ...;
    // we need to inform the container about the result
    // of the validation
    AuthenticationStatus res =
      httpMessageContext.notifyContainerAboutLogin(vr);
    return res;
  }
}
```

The secureResponse() and cleanSubject() methods are used to perform post-processing activities on requests and logout activities, respectively. Both have default implementations, so you don't have to implement them unless you need to perform additional actions on requests during the logout process.

The HttpMessageContext parameter is used as an interface to the security infrastructure controlled by the container. Its notifyContainerAboutLogin() method signals a successful login, but you can also skip the authentication process for unsecured paths via doNothing(). Or, you invoke responseUnauthorized() or responseNotFound() to force a 401 or 404 HTTP response status, respectively.

As was the case for IdentityStores, you don't have to register HttpAuthenticationMechanism implementations. You can just put them in the CLASSPATH.

In order to simplify the programming of authentication procedures, the RFC-375 standard declares a couple of variants for authentication mechanisms. They come in the form of annotations:

```
@BasicAuthenticationMechanismDefinition

@CustomFormAuthenticationMechanismDefinition(
    loginToContinue = @LoginToContinue())

@FormAuthenticationMechanismDefinition(
    loginToContinue = @LoginToContinue())
```

For example, you can use the @FormAuthenticationMechanismDefinition annotation to tell the authentication controller which URLs to use to generate the login form page and the error page, as follows:

```
@FormAuthenticationMechanismDefinition(
    loginToContinue = @LoginToContinue(
        loginPage="/login/login.xhtml",
        errorPage="/login/error.xhtml"
    )
)
@WebServlet("/secured")
@DeclareRoles({ "user", "authorized" })
@ServletSecurity(
    @HttpConstraint(rolesAllowed = "authorized"))
public class SecuredContentServlet extends HttpServlet {
    ...
}
```

To learn more about these annotations, consult the API documentation.

PART VII

Advanced Monitoring and Logging

CHAPTER 34

Monitoring Workflow

Monitoring is a vital part of any enterprise-level application. You need to be able to monitor technical aspects like memory usage, CPU load, and thread counts, performance figures like elapsed times and throughput, as well as functional aspects like sales figures, order numbers, and operation costs.

This chapter covers the process of monitoring workflows, which includes provisioning data, aggregating data, and presenting monitoring figures.

Using JMX as a Monitoring Technology

Of course it is possible to provide monitoring data by logging, by means of EJBs, web services (XML via SOAP or JSON via REST), and servlets. However, from version 1.6 on, Java includes a technology that's capable of monitoring in a more straightforward manner: JMX (Java Management Extensions). By using JMX, it is possible to declare special classes called MBeans as properties and methods accessible or callable from inside and outside the JVM. This way, applications can more easily provide monitorable data. In addition, the JVM by itself provides valuable information about its state via JMX.

MBeans come in different flavors:

- Standard MBeans

- Dynamic MBeans

- Open MBeans

- Model MBeans

- MXBeans

They differ in the way published properties or methods are reflected in the MBean coding, how they need to be registered, and which types are allowed to communicate between the client and server.

© Peter Späth 2023
P. Späth, *Pro Jakarta EE 10*, https://doi.org/10.1007/978-1-4842-8214-4_34

From the choices you have, MXBeans are extraordinarily useful for monitoring purposes. This is because communication types are limited to standard Java types with the benefit of simplified client access, and for each monitorable feature, you just need a getter method in the MBean class code. For example, consider the elapsed time in a booking workflow. All you need is an interface and an MBean class:

```
package book.jakartapro.jmxmonitoring;

public interface MonitoringBeanMXBean {
    double getElapse();
    void setElapse(double elapse);
}
```

```
package book.jakartapro.jmxmonitoring;

public class MonitoringBean
        implements MonitoringBeanMXBean {
    private double elapse;

    public double getElapse() {
        return elapse;
    }

    public void setElapse(double elapse) {
        this.elapse = elapse;
    }
}
```

In an application initialization method, you then must register the MBean with an MBean server. For simplicity, tis example uses the platform MBean server, which is configured and starts automatically for every JVM.

```
import java.lang.management.ManagementFactory;
import javax.management.ObjectName;
...
    final String OBJECT_NAME =
        "book.jakartapro.mxbeans:type=MonitoringBean";
...
    MonitoringBean m = ...;
...
```

```
try {
    ManagementFactory.getPlatformMBeanServer().
        registerMBean(m,
            new ObjectName(OBJECT_NAME));
} catch (Exception e) {
    e.printStackTrace(System.err);
}
```

Enabling Remote JMX

In order for clients to access MBeans through the network, you need to add a couple of JVM startup properties to the server:

```
-Dcom.sun.management.jmxremote=true
-Dcom.sun.management.jmxremote.port=9010
-Dcom.sun.management.jmxremote.ssl=false
-Dcom.sun.management.jmxremote.authenticate=true
-Dcom.sun.management.jmxremote.password.file=
    /path/to/jmxremote.password
```

In its simplest form, the jmxremote.password file just is a one-liner:

```
monitorRole pASSWurd
```

Of course, you should choose your own password. In the JDK-INST-DIR/conf/ management folder, you can find a template file called jmxremote.password.template that verbosely describes all the options for the password file.

Note In order to enable SSL, expect a considerable amount of extra work. You must add certificates to the server's trust store and keystore, and you must correspondingly configure your JMX clients.

Consult your Jakarta EE server documentation in order to find out how to add JVM startup properties.

417

For GlassFish, you must add a security setting to the `server.policy` file (inside glassfish/domains/domain1/config), as follows:

```
grant {
  permission javax.management.MBeanPermission
      "*", "getMBeanInfo";
  permission javax.management.MBeanPermission
      "*", "getAttribute";
};
```

MBeans in Jakarta EE Applications

Because every Jakarta EE application runs in a container-controlled environment, you can simplify the registration of MBeans to further streamline your JMX monitoring. Presuming you can use a single MBean instance for the whole application (EJB or web application), you can add the @ApplicationScoped annotation to ensure that the container takes care of constructing the MBean, and you can add a post-construct method annotated with PostConstruct to register the MBean:

```
package book.jakartapro.jmxmonitoring;

import java.lang.management.ManagementFactory;
import javax.management.ObjectName;
import jakarta.annotation.PostConstruct;
import jakarta.enterprise.context.ApplicationScoped;

@ApplicationScoped
public class MonitoringBean
      implements MonitoringBeanMXBean {
   private static final String OBJECT_NAME =
          "book.jakartapro.mxbeans:type=MonitoringBean";
   private double elapse;

   public double getElapse() {
      return elapse;
   }
```

```java
    public void setElapse(double elapse) {
        this.elapse = elapse;
    }

    @PostConstruct
    public void start() {
        try {
            ManagementFactory.getPlatformMBeanServer().
                registerMBean(this,
                    new ObjectName(OBJECT_NAME));
        } catch (Exception e) {
            e.printStackTrace(System.err);
        }
    }
}
```

Accessing the MBean for the purpose of setting monitorable values is easy then—you just use CDI to inject the MBean, as follows:

```java
package book.jakartapro.jmxmonitoring;

import jakarta.inject.Inject;
import jakarta.ws.rs.GET;
import jakarta.ws.rs.Path;
import jakarta.ws.rs.Produces;

/**
 * REST Web Service
 */
@Path("/pi")
public class Jmx {
    @Inject MonitoringBean jmxBean;

    @Produces("application/json")
    public String pi() {
        long t1 = System.currentTimeMillis();
        double pi = calcPi(100_000_001);
        long ela = System.currentTimeMillis() - t1;
```

```
        jmxBean.setElapse(ela);
        return "{\"pi\":\"" + pi + "\"}";
    }

    private double calcPi(int n) {
        // This is a stupid way of calculating PI.
        // It is just an example of a time-consuming
        // operation.
        double pi = 0.0;
        for (int i = 1; i < n; i++) {
            if (i % 2 == 1) {
                pi = pi + (4/((i*2.0)-1));
            } else {
                pi = pi - (4/((i*2.0)-1));
            }
        }
        return pi;
    }
}
```

This way, all that you need for an MBean in a Jakarta EE application is an MBean interface and an MBean class amended with a startup method.

Aggregating Values

To monitor and later show every value of a monitorable variable is not always desirable. In some cases, the amount of different numbers is just too big and keeping an eye on them becomes a tedious and error-prone task. What you want to do instead is apply some kind of aggregation, which is the process of taking, say N numbers and applying a formula to calculate a single or at least a smaller set of numbers from them.

Aggregation formulas particularly useful for monitoring are the following:

- **Moving Average**

    Takes the arithmetic average of the last N values. In order to calculate it, you use a ring-buffer of size N to gather the input values, sum up all values, and divide the sum by N. The ring-buffer ensures that the latest N values are kept in memory.

–   **Exponentially Moving Average**

Given a monitorable feature with a current value A and a parameter $\beta \in [0; 1]$, and with a new incoming value a, the next monitorable value calculates to $A = \beta \cdot A + (1 - \beta) \cdot a$. This applies smoothing with more current values having a bigger weight compared to older values. Common values for β are 0.7 ... 0.999.

–   **Standard Deviation**

Based on the same buffer as is used for the Moving Average, the *standard deviation* is calculated using the formula known from statistics. Although it's not used often, presenting the standard deviation of, for example the elapsed time, might provide valuable insight into a process' stability.

For the elapsed time example from the previous sections, a variant including the moving average reads as follows:

```java
public interface MonitoringBeanMXBean {
    double getElapse();
    void addElapse(double elapse);
}

public class MonitoringBean
        implements MonitoringBeanMXBean {
    private static final int ELAPSE_LEN = 1000;
    private double[] elapse = new double[ELAPSE_LEN];
    private int elapseInd = 0;

    @Override
    synchronized
    public double getElapse() {
        double x = 0.0;
        for(int i = 0; i < ELAPSE_LEN; i++) x += elapse[i];
        return x / ELAPSE_LEN;
    }
}
```

```
    @Override
    synchronized
    public void addElapse(double elapse) {
        this.elapse[elapseInd++] = elapse;
        if(elapseInd >= ELAPSE_LEN) elapseInd = 0;
    }
}
```

This example uses a fixed size array for the ring-buffer to make sure the code is really fast. The synchronized statements help prevent concurrent access from several threads, which could corrupt the array contents or the array index pointer value.

For a variant calculating the exponentially moving average, you don't need an array, so the MBean code is even simpler:

```
public interface MonitoringBeanMXBean {
    double getElapse();
    void addElapse(double elapse);
}

public class MonitoringBean
        implements MonitoringBeanMXBean {
    private static final double ELAPSE_EMA_PARAM = 0.95;
    private double elapseEMA = 0.0;

    @Override
    synchronized
    public double getElapse() {
        return elapseEMA;
    }

    @Override
    synchronized
    public void addElapse(double elapse) {
        this.elapseEMA =
            ELAPSE_EMA_PARAM * this.elapseEMA +
            (1.0-ELAPSE_EMA_PARAM) * elapse;
    }
}
```

Because the exponentially moving average value starts at 0.0, the algorithm needs a warm-up period before it shows valid data.

JMX Clients

There are three types of JMX clients that can access remote JMX servers:

- JMX GUI clients

- Monitoring frameworks with a JMX module

- Java applications or scripting environments using a JVM

GUI clients are programs that show JMX information of locally running or remote JMX servers. They are useful for development purposes, troubleshooting, and ad hoc monitoring needs. You normally wouldn't use them for enterprise-level operational monitoring. Examples of such applications are JConsole, VisualVM (with MBeans plugin), and some IDE plugins. This section doesn't further describe such applications, but you can readily obtain more information by consulting the corresponding user manuals.

Monitoring frameworks are programs with more elaborated features to gather, filter, store, and display monitoring data. A subsequent section covers such software in more detail.

Of course, you can also write JMX client code in the form of a Java JMX client application. Or, even better, you can use a Java scripting engine to write more concise code to access MBeans. Groovy for example enables you to write elegant JMX client code.

Note   The Groovy JMX API comes as an additional module. In order to enable it, you must add `groovy-jmx-x.x.x.jar` to the `CLASSPATH` or to the `MODULEPATH`.

The Groovy code for accessing the elapsed time from the previous sections reads as follows:

```
import groovy.jmx.GroovyMBean

import javax.management.ObjectName
import javax.management.remote.JMXConnectorFactory
    as JmxFactory
import javax.management.remote.JMXServiceURL
    as JmxUrl
import javax.management.remote.JMXConnector

def credentials = ["monitorRole","pASSWurd"] as String[]
def env = [
    (JMXConnector.CREDENTIALS): credentials
]

def serverUrl =
    'service:jmx:rmi:///jndi/rmi://localhost:9010/jmxrmi'
def server = JmxFactory.connect(
    new JmxUrl(serverUrl), env).MBeanServerConnection

def mbean = new GroovyMBean(server,
    'book.jakartapro.mxbeans:type=MonitoringBean')
println("Elapse: " + mbean.Elapse)
```

Note the capitalized E in mbean.Elapse. The field is elapse with a lowercased first letter, but GroovyMBean transforms it to a capitalized .Elapse accessor.

The power of a Groovy JMX client cannot be overemphasized. You can do lots of fancy stuff, like filtering, further aggregating, adding alarm thresholds, writing monitoring data to files or sending them as emails, presenting diagrams and charts via Swing or JavaFX, and more.

Monitoring Frameworks

Enterprise-level monitoring targets network and server availability, stability, and performance; measures technical figures describing services; and monitors application characteristics. Fully-fledged monitoring platforms usually present their activities in a web application, where diagrams and alerts are shown.

Several monitoring platform products exist. Nagios is maybe the most often cited of them, although using it in a corporate environment yields considerable license costs.[1] Another product with a friendly license and an adapter for JMX data is called Zabbix. You can download the software, including the documentation at `https://www.zabbix.com/`. Further introducing Zabbix is beyond the scope of the book, but you can get more information from the Zabbix home page.

---

[1] Nagios is a registered trademark of Nagios Enterprises, LLC. The author has no connection to Nagios Enterprises, LLC.

## CHAPTER 35

Logging Pipeline with Fluentd

From a development perspective, logging usually means including a logging library like Slf4j, Log4j or Logback, and adding logging statements to the code. Loggers then append logging information to a file, reporting functional or technical events that occur while the application is running. The logging information can be used to troubleshoot problems or for auditing purposes.

Note Depending on which library is used, loggers might also be configured to send logging data to other destinations, such as databases or message queues. However, letting loggers write to files is the most stable option if you're thinking about reliability. After all, access to the file system is provided by the operating system and there is no need to have the network running or to start additional processes on the server for logging to work. Thus, writing logging to files is by far the most commonly used method.

Beyond individual application servers and from a more enterprise-wide point of view, the need to collect logging data arises. It simplifies server maintenance and helps the operations staff do their job efficiently. For these purposes, logging data collectors can be used. They serve as an architectural component capable of collecting and aggregating logging information from multiple sources. After some optional filtering and transformation operations, the logging data can then be forwarded to different destinations, such as files, databases, monitoring applications, messaging systems, and more.

© Peter Späth 2023
P. Späth, *Pro Jakarta EE 10*, https://doi.org/10.1007/978-1-4842-8214-4_35

Collecting and aggregating logging data is not part of the Jakarta EE specification. Nevertheless, the topic is considered important enough, so this chapter talks about the Fluentd logging data collector. There are similar products, such as Logstash, Logagent, Filebeat, Rsyslog, and others, but I like the declarative way that Fluentd configures logging pipelines. In addition, Fluentd uses JSON as a message format for its internal processing and also as the dedicated output format. This makes it easy to further process logging data in central enterprise components consuming logging information.

Installing Fluentd

Fluentd can be installed on Linux, Windows, and macOS. Before actually installing Fluentd, the manual recommends a few pre-installation steps:

1. Set up an NTP daemon for accurate time stamps.

2. Increase the maximum number of file descriptors to 65535 (Linux).

3. Optimize network kernel parameters (Linux).

For details, see the documentation at `https://docs.fluentd.org/`.

Note   For testing purposes, you don't need to run any of these preparational steps.

Installation options are as follows:

- For Windows, there is an MSI installer.

- For Mac, there is a DMG package.

- For Linux, there is an RPM (Red Hat) or DEB (Debian/Ubuntu) package.

- With Ruby installed on your system, you can use `gem` to install Fluentd.

- You can compile Fluentd from the sources,

The pages at `https://docs.fluentd.org/installation` provide more details.

The Fluentd process can be run as a command or as an OS service. For simplicity, the rest of this chapter assumes that Fluentd is started using the `fluentd` command.

Running Fluentd

In order to determine whether Fluentd is running correctly, you must first create an empty folder somewhere and let Fluentd create a basic configuration inside it:

```
# terminal 1
mkdir fluentd
fluentd --setup ./fluent
```

Note   This is for Linux. For Windows or macOS, use the corresponding command syntax.

In another terminal, start Fluentd and let it run as a foreground process:

```
# terminal 2
cd fluentd
fluentd -c fluent.conf -v
```

Back in terminal 1, send input to the Fluentd process and check its output:

```
echo '{"json":"message"}' | fluent-cat debug.test
```

In terminal 2, you should now see something like this:

```
2022-03-27 10:10:22.741302646 +0200 debug.test:
    {"json":"message"}
```

In `fluentd -c fluent.conf`, you tell Fluentd to use `fluent.conf` as a configuration file. This is where the internal routing and transformation of logging messages are defined. If you change the configuration, you must restart the process (press Ctrl+C, then enter `fluentd -c fluent.conf -v` again). Of course, for your own experiments, you can use your own configuration files. Omit the `-v` flag if you want less verbose debugging output.

Using Logfiles as Input

In Fluentd, input is handled by *input plugins*. You specify them in these sections in the configuration file:

```
<source>
    ...
</source>
```

There are many types of input plugins, but for the purposes here, the most important plugin is the Tail plugin, which observes the lines added to a file. Here is a working example configuration using a logfile as input:

```
<source>
  @type tail
  path /path/to/server.log
  pos_file server.log.pos
  tag serv.log
  <parse>
    @type regexp
    expression \
        /^\[(?<logtime>[^\]]*)\] \
        (?<level>[^ ]*) \
        (?<name>[^ ]*) (?<title>.*)$/
    time_key logtime
    time_format %Y-%m-%d %H:%M:%S %z
  </parse>
</source>

<match serv.log>
  @type stdout
  @id stdout_output
</match>
```

(Remove the tailing \ characters, including the line breaks and immediately following whitespace characters.) Because of the format specification inside <parse> ... </parse>, the Tail plugin can handle input lines like these:

```
[2022-02-28 12:00:00 +0900] error
```

```
book.jakartapro.TheClass Invalid integer 42
```

The `<match>` ... `</match>` tag designates an output plugin. For now, it just prints the logging record to STDOUT.

You can save this configuration, for example, as the `conf1.conf` file, kill the Fluentd process via Ctrl+C, and restart it via `fluentd -c conf1.conf -v`.

For a simple test, replace the `path` ... line with `path server.log`, restart Fluentd, and send a few test lines to the `server.log` file in the current directory:

```
date=`date +%F' '%T' '%z`
echo "[${date}] error book.jakartapro.TheClass " \
  "Invalid integer 42" >> server.log
echo "[${date}] warn book.jakartapro.TheClass " \
  "String length > 64" >> server.log
```

The output should show something like this:

```
2022-03-28 10:55:29.000000000 +0200 serv.log:
  {"level":"error","name":"book.jakartapro.TheClass",
    "title":"Invalid integer 42"}
2022-03-28 10:55:29.000000000 +0200 serv.log:
  {"level":"warn","name":"book.jakartapro.TheClass",
    "title":"String length > 64"}
```

To give you a better idea about how this setup works, let's investigate all directives inside `<source>` ... `</source>`, step-by-step:

- `@type tail`

  Addresses the *Tail* plugin.

- `path /path/to/server.log`

  Specifies the path to the observed logfile. May contain wildcards, which allow you to observe folders with rolling logfiles.

- `pos_file server.log.pos`

  The Tail plugin needs this file to save its current state. Give it any name you like.

- `tag serv.log`

  Fluentd adds a tag to message records. This line instructs Fluentd to mark messages from the Tail plugin with `serv.log`. Tags are important, because they are used as discriminators for filters and transformers, and they determine which output plugin is used.

- `@type regexp`

  (Inside `<parse>...<parse>`.) Uses a regular expression to split each logging line into parts. Other types are described in the plugin's documentation.

- `expression ...`

  (Inside `<parse>...<parse>.`) Specifies the regular expression.

- **`time_key logtime`**

  (Inside `<parse>...<parse>`.) Specifies which part of each logging line is used to determine the log record's timestamp. Matches the `<logtime>` identifier in the regular expression.

- **`time_format ...`**

  (Inside `<parse>...<parse>`.) The format of the time string.

Filtering

Filter plugins can be used to filter out logging events based on a condition. Or you can use them to enrich and transform messages. As an example, you could amend the configuration from the previous section and add a filter for limiting Fluentd messages to the "error" logging level. To this aim, add the following to the configuration:

```
<source>
  ...
</source>

<filter serv.log>
  @type grep
  <regexp>
    key level
```

```
   pattern /error/
  </regexp>
</filter>

<match serv.log>
  ...
</match>
```

This declaration says: Allow only (by @type grep) events that have an "error" in the level field (by <regexp>).

This way, it is possible to keep the full logging in the original logfiles, while additionally sending all errors to a special destination, like a monitoring platform.

Using Multiple Routes

Message routing in Fluentd can be achieved using the output plugin called rewrite_tag_filter. The idea behind it is to send messages to different output plugins, based on a condition. The reason that this plugin is an output plugin comes from Fluentd's architecture.

Note   You must run fluent-gem install fluent-plugin-rewrite-tag-filter to install this plugin.

The following configuration lets the rewrite/routing plugin receive messages tagged with serv.log, extracts the level field, and rewrites the message using an extended tag like error.serv.log or warn.serv.log, depending on the contents of the level field:

```
<source>
  tag serv.log
  # ... must construct records with a "level" field
</source>

# Rewrite messages
<match serv.log>
  @type rewrite_tag_filter
  <rule>
```

```
    key level        # Use the "level" field
    pattern /(\w+)/ # Fetches the whole field
    tag $1.${tag}    # New tag value
    # <- The $1 comes from the pattern (the first ()-group)
    #     ${tag} contains the original tag, "serv.log"
    #     in this case
  </rule>
</match>

# Fetches the rewritten tag with level="error"
<match error.serv.log>
  @type file
  path serv.error.log
  <buffer>
    timekey 1d
    timekey_use_utc true
    timekey_wait 10s
  </buffer>
</match>

# Fetches the rewritten tag with level="warn"
<match warn.serv.log>
  @type file
  path serv.warn.log
  <buffer>
    timekey 1d
    timekey_use_utc true
    timekey_wait 10s
  </buffer>
</match>
```

This example writes its output to different files. You can of course choose from the other output plugins.

Fluentd Output

After input, message parsing, filtering, and routing, Fluentd sends the messages to its output layer. The way that messages are transmitted to the outer world is determined by the output plugins. In the previous sections of this chapter, you used the `stdout` plugin to send messages to the console, and the `file` plugin wrote the message records to local files.

There are many more output plugins . An almost complete list of them is as follows:

- `null`

  Drops messages.

- `stdout`

  Writes messages to the console attached to the Fluentd process. You use this plugin mainly during development and for debugging purposes.

- `forward`

  Sends messages to other Fluentd nodes. The plugin is very elaborate and there are many options to control the message flow.

- `file`

  Writes to the local file system. Writes to rolling files based on a freely configurable time period.

- `exec`

  Sends messages to an operating system command. This plugin is actually quite powerful, since you can basically do anything you like with logging records. Even if the starting up operating system processes are slow compared to built-in output plugins, you still can use the `exec` plugin efficiently in a heavy load environment, because you can use buffers to collect message records in memory, and then send messages in bulk to the OS process.

- `http`

  Sends messages to an HTTP server. This could be a REST interface receiving JSON formatted data.

- mongo and mongo_replset

  Sends messages to a MongoDB database or MongoDB replica set.

- s3

  Forwards data to an Amazon S3 cloud storage instance.

- webhdfs

  Sends messages to an Hadoop instance.

- kafka

  Sends messages to Apache Kafka for further message routing.

- elasticsearch

  Sends messages to ElasticSearch.

- opensearch

  Sends messages to OpenSearch.

Five more output plugins—copy, roundrobin, exec_filter, relabel, and rewrite_tag_filter—do not actually output data, but instead control the routing of data in Fluentd. You can read more about those plugins in Fluentd's documentation.

Further Reading

This chapter covered only a subset of Fluentd's capabilities. The best place to learn more about Fluentd is from the official documentation at https://docs.fluentd.org/.

CHAPTER 36

Performance Troubleshooting

In many IT projects, application performance is monitored too late in the project timeline. Sometimes application performance is completely neglected, which too often results in unsatisfied customers and expensive software fixes while the application runs in a production stage. It is therefore advisable to include performance tests and performance optimization early on in a project, such as during development, or at least during the test phase.

This chapter investigates techniques and best practices for optimizing application performance.

Note The ideas presented in this chapter are not limited to Jakarta EE applications. You can use them for other server technologies as well.

Load and Performance Tests

In order to see how your application will behave in real-world scenarios before it goes to production, you must add a testing phase to your project plan. Testing is a science of its own, but there is a distinction between unit tests, integration tests, functional test, and nonfunctional requirement (NFR) tests. While unit tests cover components as they are written by developers, functional tests focus on behavioral aspects of an application, and integration tests try to determine whether the various components of an application work together well in an IT landscape. Finally, NFR tests check the performance of an application or the application components, and they also examine the stability of an application under load (i.e., with many concurrently acting, simulated users at work).

437

© Peter Späth 2023
P. Späth, *Pro Jakarta EE 10*, https://doi.org/10.1007/978-1-4842-8214-4_36

Consider for example a web page that books a service. In a corresponding test, you should check the following:

- When data is entered and submitted, are subsequent business processes—such as shipping, allocating business resources, and billing—triggered? Does the next page loaded show the expected result? Is the database updated correctly? This refers to functional and integration tests.

- What is the elapsed time between submitting the data and loading the next page (such as a confirmation page)? This refers to performance tests. For more realistic results, you should run these tests while the application is under load. This means you simulate many, concurrently acting users.

- When the application is under load, does it still work correctly? Are enough computational resources available? This refers to load tests.

- When NFR tests are running, how does the memory consumption look? What can you say about the process load as seen from the operating system? Such secondary data can improve the value of NFR tests, and it would be nice if an NFR test tool could add such figures.

Performance and load tests are not limited to web pages by talking to web components using HTTP. You can also directly address web services (SOAP) or REST interfaces. You can even test mail servers by writing NFR tests that communicate via SMTP, or you can write tests targeting JMS endpoints. This flexibility allows for a fine-grained test setup, which in contrast to coarse tests, simplifies the identification of bottlenecks.

NFR Testing Methodology

An NFR test concept contains the following elements:

- Identify crucial application components with a web, web service (SOAP), or REST interface. Any other interface is possible as well, but these three are the most important ones.

- Develop test cases. A test case is a series of test steps with the following characteristics:

- A test case simulates a business activity, like an order or a user registration.

- A test step is an atomic action, like sending a GET or POST request to a web application or to a REST interface.

- A test case might contain preparational steps. It must be possible to automate such preparational steps. For example, if the test case describes the deletion of a user from the user database, adding the user is a preparational step.

- A test case might contain cleanup steps. It must be possible to automate such cleanup steps. For example, if the test case describes the registration of a user with the user database, deleting the user afterward is a cleanup step.

- A test case must be repeatable.

- A test case must be parallelizable. This means it must be possible for several threads to perform the same test case at the same time, without interfering with each other.

- Describe a tool capable of running test cases. This chapter uses JMeter for this purpose.

- If a test case simulates a web application workflow, decide whether frontend (web browser) activities will be included. If this is the case, use Selenium as a testing tool, since JMeter cannot simulate browsers.

- Decide whether monitorable data needs to be added. This could, for example, be memory consumption as seen by the JVM or by the operating system.

- Describe the analysis of NFR test outcome data.

- Suggest improvements in order to increase performance or stability.

A collection of test cases, executed sequentially or in parallel, is sometimes called a *test plan*. JMeter uses this idiom.

Where to Run NFR Tests

Test plans can be developed on a user workstation. In many cases, it is also possible to run NFR tests on developer machines. Using dedicated servers for the same purpose is an option too. Experience shows that both approaches are valid, but the following checklist should help you to make your decision:

- If user workstations experience high network latency when connecting to the application server, a dedicated NFR test server is the better option.

- NFR tests can run for hours. If the user workstations' network connection is not 100 percent reliable, a dedicated NFR test server is the better option.

- Developer machines usually have enough power to simulate 30 or even 100 concurrently acting users. If this is not the case or if you need the PC to also run other applications that require significant CPU resources, a dedicated NFR test server is the better option.

- Using user workstations for NFR tests is sometimes more realistic compared to a dedicated NFR test server. In the end, application clients are often user workstations. If this is the case for you, running NFR tests on a developer machine is the better option.

- Using a dedicated NFR server means you have to transport JMeter test plans from the developer PC to the test server, before you can start the tests. This is not rocket science, but it takes a few minutes each time.

- If you need to include Selenium with your NFR tests, using a dedicated NFR test server usually is not an option.

- Using developer PCs for short-term tests and dedicated NFR test servers for longer running tests is an option as well. This setup gives you the maximum amount of flexibility.

Caution   Never run NFR tests on the same server as the application. Such a setup generates weird effects because of back-coupling.

NFR Test Duration

While you might be tempted to schedule NFR tests for less than an hour or so, experience indicates that it is advisable to include longer-running test in your project plan. Running NFR tests overnight for six to ten hours provides performance data with better statistical significance. Also, deficiencies like memory and classloader leaks often show up only after a couple of hours.

The suggested procedure is to mix tests that run an hour or two, perhaps during business hours, with tests that run overnight, for six to ten hours.

NFR Tests with JMeter

JMeter is an extremely versatile tool for creating and running NFR test plans. It has a GUI for building tests of any level of complexity. As you add more and more test elements, you can run the tests from the GUI and investigate test outcomes using real-time diagrams or automatically updated tables. Later, when the test plan is finished, you can run it from the terminal on the developer machine or after you copy the test plan to a dedicated NFR test server.

Since JMeter is a Java tool, you can freely choose which operating system you want to use. You can, for example, develop test plans on a Windows PC and later run them on a Linux server.

As a very helpful feature, JMeter provides a recorder for web applications. You can configure your browser to send its requests through a proxy provided by JMeter. All requests, including redirections, then show up as test elements in a generated test plan. This greatly simplifies and speeds up test-plan development for complex web application workflows.

Note   This chapter doesn't cover this recording feature in detail, since it is beyond the scope of the book. Visit `https://jmeter.apache.org/usermanual/component_reference.html#HTTP(S)_Test_Script_Recorder` to learn more about it.

You can download JMeter from `https://jmeter.apache.org/`. This is also your primary place for getting information.

In order to simplify the process of starting up JMeter, add the `start.sh` file to the installation folder with these contents:

```
#!/bin/bash

# Add your own JDK path here
export JAVA_HOME=/opt/openjdk-17.0.1

# Set English as GUI language, and add module
# setting for JDK >= 9
export JVM_ARGS="-Duser.language=en
  --add-opens java.desktop/sun.awt.X11=ALL-UNNAMED"

# Launch JMeter
bin/jmeter.sh
```

For Windows, you add a corresponding file, called `start.bat`:

```
set JAVA_HOME=C:\java\openjdk-17.0.1

set "JVM_ARGS=-Duser.language=en ^
  --add-opens java.desktop/sun.awt.X11=ALL-UNNAMED"

bin\jmeter.bat
```

To get started, you'll see how to develop a simple test plan for loading the index page of a web application. Figure 36-1 shows JMeter's start window (choose Options ➤ Look and Feel and select your favorite color scheme).

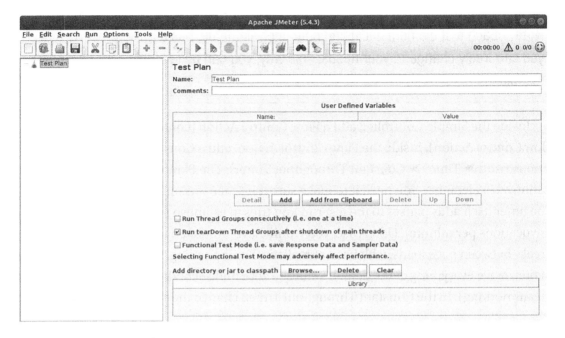

Figure 36-1. *JMeter's start window*

Right-mouse-click Test Plan in the left tree view panel, then choose Add ➤ Threads (Users) ➤ Thread Group. This is a master controller where you specify the parallelization grade (the number of simulated concurrent users) and the test lifetime. In the settings view for the thread group, make these changes:

```
Number of Threads (users):   50
Ramp-up period (seconds):    30
Loop Count: [x] Infinite
[x] Specify Thread lifetime
Duration (seconds):          300
```

Leave the other settings at their default value. This defines a test running for five minutes, simulating 50 users, and a ramp-up period of 30 seconds. The last option indicates that the users will not all start their work at once, which is unrealistic. Instead the startup time for the users is evenly distributed among the first 30 seconds.

Inside the thread group, add a Simple Controller (choose Add ➤ Logic Controller ➤ Simple Controller). This is simply a container for other test elements; it helps you organize the test plan.

Note   Each test element, including container elements, has a name property that you can freely change to your needs. This way, you can build highly comprehensive test plans.

Inside the Simple Controller, add a Flow Control Action (choose Add ➤ Sampler ➤ Flow Control Action). Inside the Flow Control Action, add a Constant Throughput Timer (choose Add ➤ Timer ➤ Constant Throughput Timer). The Flow Control Action is just a dummy test element—you need it to make sure the timer acts only once each iteration. The timer itself adds pauses to the test run and tries to establish a certain throughput (invocations per minute). Otherwise, you would have simulated users who act without breaks between page loads, which is not realistic. For the setting, inside the Flow Control Action, leave everything at the default values (a zero milliseconds pause, which basically means nothing). In the Constant Throughput Timer, change the settings as follows:

```
Target throughput (per minute): 6.0
Calculate based on:             this thread only
```

This simulates each user clicking the Reload button every ten seconds.

As a basic test plan element, add a HTTP request to the Simple Controller (as a sibling to the Flow Control Action). You can find it by choosing Add ➤ Sampler ➤ HTTP Request. In its settings page, change the following:

```
Server Name or IP: www.someserver.com
Port Number:       80
Path:              /index.html
```

Use any running web server of your choice here. However, do not use a locally running server—the response times are just too slow.

You still need a way to save or display the test outcome data. This is covered by *listeners*. You'll add three listeners to the test plan:

- A Summary Report shows performance data in a table, updated in real-time while the test is running.

- A Graph Results listener shows a quick diagram with performance figures. This diagram is updated in real-time, too.

- A Simple Data Writer writes performance figures to a CSV file.

For all three listeners, only a filename for the Simple Data Writer needs to be entered. You can play around with the other settings at will. I place the listeners as siblings to the thread group.

The test plan now looks like Figure 36-2. You can start the test from the JMeter GUI. Select the Graph Result listener in order to see how the performance figures develop.

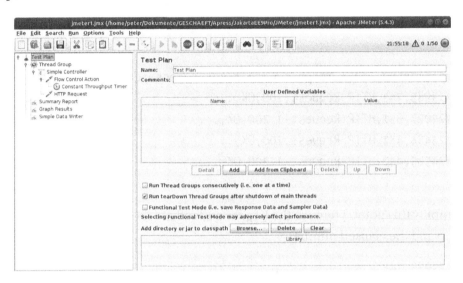

Figure 36-2. *JMeter test plan*

Frontend Tests with Selenium

Selenium is a browser automation tool. Contrary to JMeter, Selenium simulates real browser activities, including JavaScript code. There are two operating modes for Selenium: either you use a web driver scriptable via Java API (other languages are available), or you use the Selenium IDE, which is a browser plugin with GUI-style scripting capabilities.

Although technically possible, you normally don't use Selenium to generate significant load on a web application server. Running many Selenium instances in parallel requires the allocation of a huge amount of CPU resources. And you don't need that—in the end, server applications are not getting triggered by client applications, but by the protocol data clients send to servers. If you really think Selenium will help you reach your project goals, it makes much more sense to generate load via JMeter and then let just *one* Selenium instance run in parallel.

You can download Selenium from `https://www.selenium.dev`. This is also where you can find other documentation.

Analyzing Performance Figures

After you run a test, the elapsed time of the whole workflow or some part of it is the most interesting outcome. The output from JMeter's Simple Data Writer listener could look as follows:

```
timeStamp,elapsed,label,responseCode,responseMessage,...
1649233423985,743,HTTP Request,200,OK,...
1649233424077,651,HTTP Request-1,200,OK,...
1649233424484,481,HTTP Request,200,OK,...
1649233424525,440,HTTP Request-1,200,OK,...
...
```

To simplify this point, I remove the header and reduce the data set to just the timestamp and the elapsed number:

```
1649233423985,743
1649233424077,651
1649233424484,481
1649233424525,440
...
```

The Linux command to perform this transformation is as follows:

```
cat testout.csv | tail -n+2 | cut -d',' -f1,2 \
    > testout2.csv
```

Reducing the File Size

After a long test, the test outcome file may contain 100,000 or more lines. Handling such big files inside a spreadsheet program, maybe to draw diagrams, is a very time-consuming and tedious task. For this reason, you can reduce the file. Keeping only every Nth line and throwing away all the other lines is easy. Just enter the following in a terminal to keeping every second, third, or forth line:

```
cat testout2.csv | awk '(NR+1)%2==0' > testout3.csv
  - or -
cat testout2.csv | awk '(NR+1)%3==0' > testout3.csv
  - or -
cat testout2.csv | awk '(NR+1)%4==0' > testout3.csv
...
```

Plotting a Performance Chart

You can now import your data into a spreadsheet program, for example LibreOffice Calc. After the import, you get the timestamps in Column A and the elapsed time in Column B. As a first refinement, you can add a column for the test timeline. To this aim, insert a new column next to the first column and enter this formula inside cell B1:

```
=(A1-A$1)/1000
```

Use the fill handle of cell B1 to fill the whole column with this formula; see Figure 36-3.

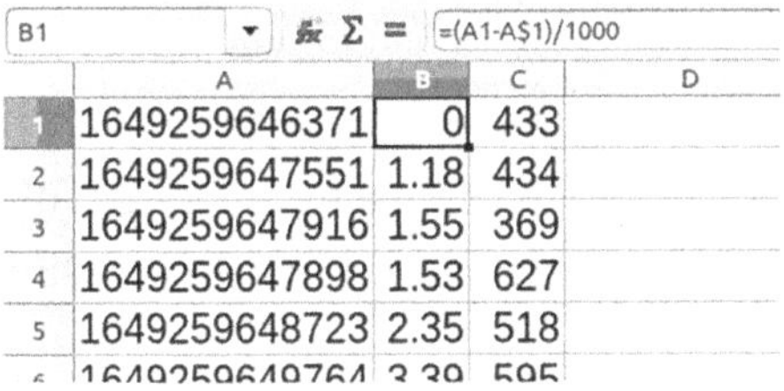

B1		f_x Σ =	=(A1-A$1)/1000	
	A	B	C	D
1	1649259646371	0	433	
2	1649259647551	1.18	434	
3	1649259647916	1.55	369	
4	1649259647898	1.53	627	
5	1649259648723	2.35	518	
6	1649259649764	3.39	595	

Figure 36-3. *Test timeline has been added*

Note   Press Ctrl and Left+Mouse-Double-click the fill handle to fill the whole column at once.

If you now create a chart from the raw data, you'll get a cloud like the one shown in Figure 36-4.

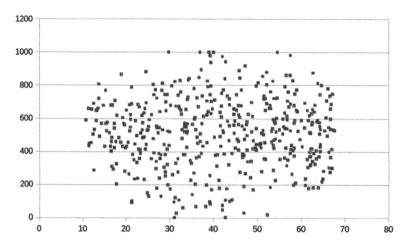

Figure 36-4. *Elapsed time chart*

This chart is not very expressive—you cannot reliably tell what the average elapsed time is, nor how the elapsed time develops during a test. For this reason, you can add another column and calculate the moving average. Inside the cell at D20 add this formula:

=SUM(C1:C20)/20

Fill all cells below cell D20 using the fill handle of cell D20. Figure 36-5 shows the result. This produces a moving average using a frame length of 20 items. Using ten items instead (change the formula!) generates a less smooth line—bigger frames lead to more smoothing.

	A	B	C	D
11	1649259651574	5.2	365	
12	1649259652907	6.54	514	
13	1649259652905	6.53	648	
14	1649259653187	6.82	410	
15	1649259653457	7.09	439	
16	1649259653348	7	610	
17	1649259653450	7.08	523	
18	1649259654132	7.76	572	
19	1649259654707	8.34	367	
20	1649259654528	8.16	552	494.05
21	1649259654840	8.47	397	492.25
22	1649259655281	8.91	521	496.6

D20 | fx Σ = | =SUM(C1:C20)/20

Figure 36-5. *Moving average column added*

Another interesting figure is standard deviation. It tells you how much the elapsed times differ from their average. The formula for standard deviation is as follows:

$$\sqrt{\frac{\sum_{i}^{n}\left(y_i - \bar{y}\right)^2}{n-1}}$$

To add it to the spread sheet, you first use a column for the squared distance from the elapsed time to its moving average. In cell E20, write the following:

```
=(D20-C20)*(D20-C20)
```

Fill all cells below cell E20 using the fill handle of cell E20. Now, in cell F30, you can add the formula for the standard deviation:

```
=SQRT(SUM(E21:E30)/9)
```

Again, use the fill handle to copy this formula to all cells below F30.

You can now plot a chart from columns B, C, D, and F. It will look like Figure 36-6.

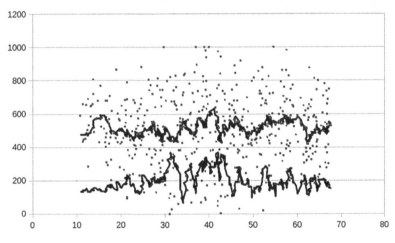

Figure 36-6. *Moving average column added*

No Min or Max, Please

It's customary to watch for minimum and maximum values in time series. However, the information behind such numbers is of negligible worth. Consider, for example, a time series of elapsed times with 999 values near 500ms and one value at 3.0s. The max value is obviously 3.0s. If you take a second time series with 500 values near 500ms, and 500 values between 2.5s and 3.0s, the max value again is 3.0s. But the two time series are totally different! There is no value in stating that there is a maximum of 3.0 seconds.

Much better information about the distribution of values is given by the *percentiles*. The calculation is as follows: given N values $y_i (i = 1, ..., N)$, the M-Percentile (M between 0 and 100) is given by exactly the y_j, where M percent of all the other y_i are smaller than y_j. Computationally, you sort the y_i in ascending order and then fetch the row at index $N \cdot M/100$.

For example, if you have $N = 1000$ for 1000 values y_i, you first sort the y_i in ascending order. For the 95 percentile, you look up the value at row number 950 (from $1,000 \cdot 95/100$). For example, if this value is 650ms, you can say: "The 95 percentile of the series lies at 650ms," which is equivalent to "95 percent of all values are below 650ms."

Common percentiles are 90 percentile, 95 percentile, 98 percentile, and 99 percentile. Since percentiles are easy to calculate, a performance test report would probably contain all of them.

Code-Level Monitoring with VisualVM

VisualVM is a monitoring and management tool that addresses various aspects of a running JVM. It has a built-in plugin called Sampler, which queries the JVM state on a regular basis, including elapsed times of method calls. This way, it is possible to find performance bottlenecks without writing a single line of code or changing the server configuration.

Note   You can download VisualVM at `https://visualvm.github.io/index.html`.

VisualVM can be run with a Java version different from the application server. There is a minor issue with VisualVM and OpenJDK 16 or 1—the tooltip texts do not reliably show up, which makes it a little bit hard to handle. With OpenJDK 15, there is no problem.

As an example, you could add a time-consuming operation to a REST controller. The following method shows a somewhat unclever way of calculating the `pi` constant:

```
@Path("/pi")
@GET
```

```
@Produces("application/json")
public String pi() {
  return "{\"pi\":" + calcPi() + "}";
}

private double calcPi() {
  int d = 1;
  double sum = 0;
  for(int i=0;i<100_000_000;i++) {
    if(i % 2 == 0) sum += 4.0/d; else sum -= 4.0/d;
    d += 2;
  }
  return sum;
}
```

You can write a new web application containing the pi() method, or you can temporarily add it to an existing controller—it is up to you.

In a JMeter test plan, you simulate many concurrently acting users calling the pi() method. For simplicity, you can use an infinite loop for this purpose.

If you start VisualVM, connect it to the GlassFish application, and open the Sampler plugin, you can click the CPU button to start fetching performance data. Click the Hot Spots view button. After a few seconds, you will be presented with a table showing the most time-consuming methods. See Figure 36-7.

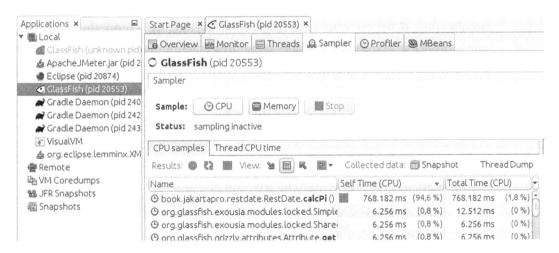

Figure 36-7. *VisualVM at work*

As expected, the `calcPi()` shows up on top of this list.

Code Optimization

There is a plethora of possible reasons for performance bottlenecks. It is not possible to construct something like a checklist in order to avoid all of them. Nevertheless, the following list gives you a couple of hints to avoid such shortcomings:

- Depending on the circumstances, functional constructs may be superior compared to procedural setups. However, you must watch out for intermediate terminal operations. Building large collections—like lists, sets, or maps—in a functional expression is a time-consuming operation.

- Avoid the construction of too many objects in large loops.

- Although often considered old fashioned, using arrays instead of collections can speed up an application significantly.

- For database operations, consider using a cache.

- Unless absolutely necessary, avoid using reflection. Reflective access to Java objects is much more time consuming compared to direct, language-level access.

- Using JMS to transmit data often leads to a clean and stable architecture. However, watch out for pitfalls—messaging is more time consuming compared to more directly addressing remote objects, for example via REST interfaces or via RMI. On the other hand, due to its asynchronicity, JMS messaging might improve performance. A detailed analysis can help avoid deficiencies.

- Avoid using JMS for concurrent constructs. The current versions of Jakarta EE handle concurrency much more efficiently than older versions, where JMS was abused for the same purpose.

- When possible, use JSON over XML. JSON is leaner than XML and processing large JSON structures is less time consuming than for equivalent XML data.

- Avoid file operations, when possible. Writing data into memory instead of into files is much faster.

- Avoid overly extensive logging.

- An ill-configured garbage collection program might lead to performance problems. Chapter 37 talks about garbage collection.

Garbage Collection

Garbage collection is the cleanup process that Java uses to get rid of unused objects, freeing associated heap memory. Over time, Java garbage collectors went through many changes, and the historical development of garbage collectors is actually quite interesting. However, a detailed description of ancient garbage collectors is beyond the scope of this book, so this chapter limits its description to the three garbage collectors most often in use—The G1 Garbage Collector, the Shenandoah GC, and the Zero Garbage Collector. The inclined reader can find information about the other garbage collectors on the web.

The Importance of Garbage Collectors

Garbage collection happens in the background, while the application server is running. From an application developer's point of view, garbage collectors do their work without interfering with the application—it is not possible for garbage collectors to do something functionally interesting. Nevertheless, garbage collectors (GCs) can influence the application's operation in one or a combination of ways:

- If the GC is wrongly configured, it might fail to properly do its work. Unfreed memory might pile up, eventually leading to an out-of-memory error and an application breakdown.

- GCs need one or more threads to do their work. This requires computation power that the application cannot use.

- Due to the way GCs function, GCs have to stop the application threads once in a while. Obviously, such STW (stop the world) periods should be reduced to a minimum. The STW period length is often referred to as *latency*.

© Peter Späth 2023
P. Späth, *Pro Jakarta EE 10*, https://doi.org/10.1007/978-1-4842-8214-4_37

- A bug in the application might prevent the GC from freeing up unused objects. This can cause an application to break down. Such bugs are called *memory leaks*.

For these reasons, GC figures should be included in performance/stability tests. Fortunately, this is not very complicated. All GCs publish their own performance figures via JMX. In JMeter, for example, you can add Groovy scripts that write numbers queried via JMX into other test outcome files.

There are no rules for determining which GC best suits your application, and which configuration options to choose for maximum GC performance and stability. For this reason, you must follow a heuristic approach and run performance and load tests with various GC options in order to see which GC setup is the best for you.

G1 Garbage Collector

The G1 garbage collector, also called *garbage first*, was designed with large heap sizes and multi-processor machines in mind. It is the default GC since JDK version 9.

Enter `java g1 options` in your favorite search engine to find out more about G1 options.

Shenandoah GC

The Shenandoah GC is designed for lower latencies compared to G1, especially for larger heaps. It was optionally included in JDK 11, but needed an opt-in during build time. It was discontinued for JDK versions 12 through 16, but came back fully included within JDK 17. You can switch on this garbage collector via this flag:

```
-XX:+UseShenandoahGC
```

Enter `java shenandoah options` in your favorite search engine to find out more about Shenandoah GC options.

Zero Garbage Collector

The Z Garbage Collector is a low-latency garbage collector with STW (stop the world, i.e. pausing the application) periods guaranteed to be less than ten milliseconds.

It also claims to handle terabyte heaps without significant degradation. For heavy-traffic applications with maybe 1,000 or more users, the Zero Garbage Collector might be your best choice.

To enable the Zero Garbage Collector, add the following to the java command:

```
-XX:+UseZGC
```

This is possible in JDK 15 or later. For JDK 11 to 14, the Zero Garbage Collector was an experimental feature and you had to also add:

```
# only for jdks < 15
-XX:+UnlockExperimentalVMOptions
```

Enter java zero garbage collector options in your favorite search engine to find out more about ZGC options.

Garbage Collector Logs

For all garbage collectors, you can enable logging output. The garbage collector then writes information about its activities into a dedicated logfile. To analyze such files is a science of its own, but besides writing scripts to extract valuable information from GC logfiles, there are also GUI applications that can help you. For example, the little tool GCViewer from https://sourceforge.net/projects/gcviewer/ can help you quickly analyze GC activities.

Note   For garbage collector logging options, watch out for JVM -Xlog:* or -Xlog:gc:* flags in the documentation (* is a wildcard).

CHAPTER 38

Memory Troubleshooting

A central characteristic of Java programs is the dynamic creation of objects while the application is running. In well-formed code, the garbage collector can automatically remove unused objects in a balanced manner. That is to say that, at some point, the number of removed objects equals the number of newly created objects. Unfortunately, and especially with complex applications in a corporate environment, software deficiencies might lead to functionally unused objects somewhat hanging around and *not* being cleared by garbage collection. If such objects pile up over time, that's called a *memory leak*.

Memory leaks inevitably lead to an application breaking down. It is therefore important to identify memory leaks during the testing phase, identify their cause, and fix the code.

Identifying Memory Leaks

Consider the following code snippet:

```java
static class A {
    int x = 7;
}

static List<A> list = new ArrayList<>();

public String memLeak() {
    Runnable r = new Runnable() {
        public void run() {
            while (true) {
                for (int i = 0; i < 500; i++) {
                    A a = new A();
```

© Peter Späth 2023
P. Späth, *Pro Jakarta EE 10*, https://doi.org/10.1007/978-1-4842-8214-4_38

```
                    if (Math.random() < 0.5)
                        list.add(a);
                }
                try {
                    Thread.sleep(0, 100);
                } catch (InterruptedException e) {
                }
            }
        }
    };
    new Thread(r).start();
    return "{\"result\":\"" + "OK" + "\"}";
}
```

You can for example add this to a REST controller inside a web application:

```
@Path("/")
public class MemLeak {
    static class A {  int x = 7;  }
    static List<A> list = new ArrayList<>();

    @Path("/memleak")
    @GET
    @Produces("application/json")
    public String memLeak() {

        ...

    }
}
```

Once you invoke the method, it spawns a new thread. Inside the thread, the program enters an infinite loop for creating instances of class A. Each object is added to a list with a probability of approximately 50 percent. The garbage collector later removes the objects not added to the list. The objects inside the list can never be freed, and they perpetually pile up in the heap memory. So you effectively have an artificial memory leak here.

If you added the method to a REST controller, the application happens to be called MemLeak, and the servlet filter addresses webapi/* URLs (configured inside web.xml). You can start the leaking method via this:

```
curl -X GET http://localhost:8080/MemLeak/webapi/memleak
```

Note   This is for a locally running GlassFish—for other application servers, you probably have to change this URL.

In a VisualVM instance connected to the server process, you can watch the heap memory usage from the Monitor tab. With the leaking method running, you will see a development like the one shown in Figure 38-1.

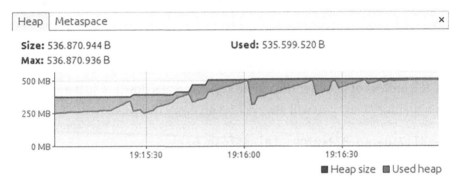

Figure 38-1. Memory leaking

You can see that the Used Heap figure hits the maximum allowed heap space (512 MB in this case) several times. Garbage collector activities, recognizable by negative slope regions, free less and less memory. Finally, the Used Heap curve approaches the maximum heap value and the garbage collector can't free any noticeable amount of memory. This is the point where the application server finally breaks down.

Unfortunately, plotting the Used Heap line will not always point to the leaking application. Consider another example shown in Figure 38-2.

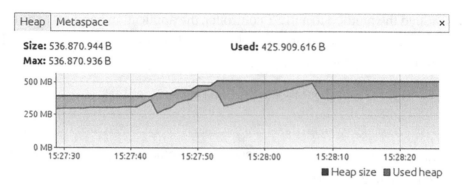

Figure 38-2. *Another memory leaking*

Here, the Used Heap number touches the maximum value just once, so the garbage collector performs a cleanup, and the application server continues its work with a flatter slope with respect to the Used Heap number. The leaking thread dies, not the whole application.

Looking at these two examples, it should be clear that you need another source of information in order to reliably detect memory leaks. This is why you also need to look at the logs. Regardless of whether the whole server breaks down or just an application thread, for a memory leak, you will always find an entry like this:

```
[2022-04-14T17:06:40.873+0200] [glassfish 6.2]
[SEVERE] [] [] [tid: _ThreadID=305 _ThreadName=Thread-8]
[timeMillis: 1649948800873] [levelValue: 1000] [[
  java.lang.OutOfMemoryError: Java heap space
      at ...
      at book.jakartapro.memleak.MemLeak$1.run(
          MemLeak.java:32)
      at java.base/java.lang.Thread.run(Thread.java:834)
]]
```

The suggested procedure is to first look at the logs, and then visualize the Used Heap curve to strengthen the evidence for a memory leak in action.

More Evidence: Heap Dumps

To further identify which part of an application is responsible for a memory leak, you can use *heap dumps*. Heap dumps represent the precise memory state a Java application (or application server) has at the time the heap dump was taken. A heap dump stores its data in a file, typically located on the same machine on which the application runs.

There are several possibilities for taking heap dumps. Some of them require the knowledge of the process ID of the Java program you want to take the heap dump from.

Note   For all the shell commands presented here, you must either add the `bin` directory from your JDK installation to the PATH environment variable, or you must add the full path to the command.

In order to acquire the process ID, use the `jps` command:

```
jps -l
```

```
Output for example:
8384 jdk.jcmd/sun.tools.jps.Jps
8209 org.eclipse.lemminx.XMLServerLauncher
7796 /opt/eclipse/[...]equinox.launcher[...].jar
7416 com.sun.enterprise.glassfish.bootstrap.ASMain
8072 org.gradle.launcher.daemon.bootstrap.GradleDaemon
```

In the output shown, `ASMain` points to a running GlassFish instance—7416 is the process ID in this case.

Back to the heap dumping tools—the `jmap` command is one option:

```
jmap -dump:live,format=b,file=dump1.hprof 7416
```

The `live` flag is optional. If you omit it, orphaned objects subject to garbage collection will be added to the dump. The `format=b` flag ensures that the dump will be in HPROF binary format. 7416 is the process ID; substitute your own here. After the command has executed, you can find the dump in the current directory.

The jmap command has an operation mode where just the Java object instances are counted and no dump is written. Enter the following:

```
jmap -histo:live 7416 | less
```

To see output like this (which has been slightly shortened):

```
num #instances #bytes     class name (module)
-------------------------------------------------------
1:   7909999   126559984  book.jakartapro.memleak.MemLeak$A
2:    179397    45847984  [Ljava.lang.Object; (java.base@11)
3:    403798    42437568  [B (java.base@11)
4:    128053    11268664  java.l.r.Method (java.base@11)
5:    388874     9332976  java.l.String (java.base@11)
6:     12250     5787624  [I (java.base@11)
...
```

This might help find memory leak causes. For this example, you can see there are suspiciously many instances of the MemLeak$A class, as expected. Additionally investigating the full dump might still be necessary, though. Only the full dump allows for investigating object relations.

Another heap dumping tool is called jcmd. Use it as follows:

```
jcmd 7416 GC.heap_dump dump1.hprof
```

The first argument is the process ID of the application server. The dump file is *not* created in the current directory. Instead, you have to look for it in the application server's working directory. For GlassFish, this is GLASSFISH_INST_DIR/glassfish/domains/ domain1/config.

Heap dumps can also be taken from inside the VisualVM tool. After connecting VisualVM to the application server instance, you can find a Heap Dump button on the Monitor tab. Once the heap dump is taken, VisualVM opens an extra tab with useful information on it, as shown in Figure 38-3.

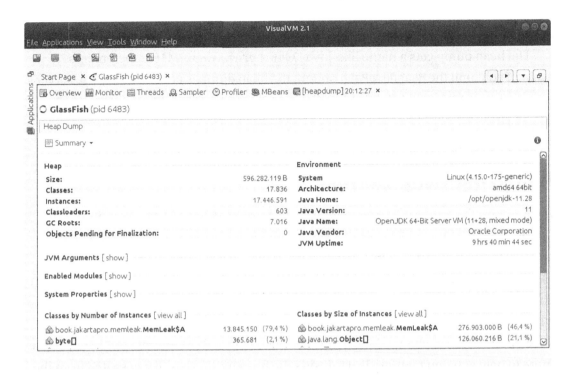

Figure 38-3. VisualVM heap dump

VisualVM saves heaps in a temporary place. To save a heap dump in a file for later analysis, right-click the heap dump in the tree view on the left side of the window and select Save As from the menu.

The JDK contains a GUI tool called `jconsole`, which you can also use to trigger a heap dump. After you start it, a connection dialog box allows you to select the server process. Navigate to the MBeans tab and select the `com.sun.management.HotSpotDiagnostic` MBean. Inside its Operations view, there is a `dumpHeap` method, which you can invoke using the appropriately named button. Use the filename for the parameter, for example `heap.hprof`, and use `true` to start the garbage collection. After the heap dump creation is triggered, you can find it in the application server's working directory. For GlassFish, this is `GLASSFISH_INST_DIR/glassfish/domains/domain1/config`.

As a last option, you can add a JVM startup option that signals the process to automatically generate a heap dump before an `OutOfMemoryError` is thrown. To this aim, add the following flags to the startup script:

```
-XX:+HeapDumpOnOutOfMemoryError
```

```
-XX:HeapDumpPath=<file-path>
```

The heap dump gets a name like `java_7416.hprof`, where 7416 is the process ID again. If you omit the `HeapDumpPath` option, the heap dump is created inside the server's working directory (`GLASSFISH_INST_DIR/glassfish/domains/domain1/config` for GlassFish).

Analyzing Heap Dumps

Heap dumps use a binary format named `HPROF` to store class and object relation information. Besides, for a Jakarta EE server, you typically have thousands of classes at work. So you can imagine that analyzing a heap, or comparing two heaps taken at different times, is a very challenging task. Fortunately, there are a couple of tools to help you here—Eclipse Memory Analyzer and VisualVM. Both can be used without acquiring a license.

You already read about taking heaps from inside VisualVM. But the program can do more to help you find memory leaks. Inside the heap dump view, if you switch from the Summary to the Objects view type, you enter a tool subset capable of telling you about instance statistics and object relations. See Figure 38-4.

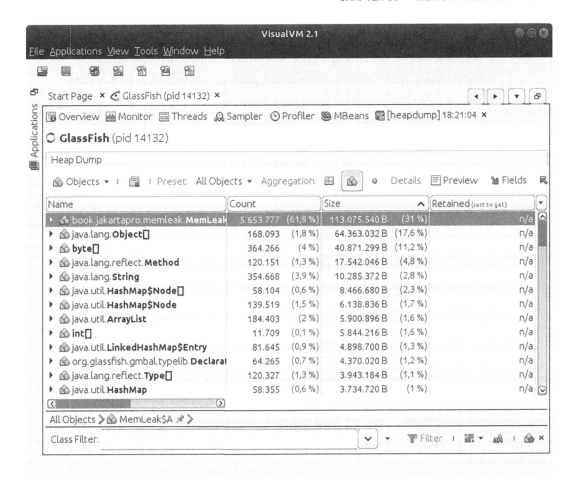

Figure 38-4. *VisualVM object view*

In the checkbox labeled Preset, you can select object relations you are interested in. Possible values are as follows:

- **All Objects**

    Look at all objects, regardless of what their relations to other objects are. In this preset, you can determine that there are exceedingly many instances of the MemLeak$A class. You cannot, however, find out where the objects are stored, or why the objects are not being garbage collected. This preset thus only gives you a hint as to which class is the source of the memory leak.

- **Dominators**

  The *Dominator* relationship expresses which objects are responsible for preventing other objects from being garbage collected. A Dominator view is a tree view, because you can arrange Dominators in a tree hierarchy. It allows you to calculate *retained* size, which is the number of bytes that can be freed if the dominating relationship is broken. The Dominators preset thus is very important for memory leak-related troubleshooting.

- **GC Roots**

  Look at GC roots, which are the primary classes of a JVM from which all relations are built. Garbage collection algorithms start at GC roots. Because, from an application point of view, GC roots have a more technical semblance, you typically don't look at GC roots to analyze memory leaks.

The Aggregation type allows you to select the aggregation mode applied to the objects, and in Details, you can trigger various calculations for a selected item in the list.

If you use the Dominators preset, select the Instance aggregation and sort by Retained (click the column header). You can then see that the single `MemLeak$A` instance retains a big part of the heap and is dominated by the `list` static field. Voila—there is the memory leak! See Figure 38-5.

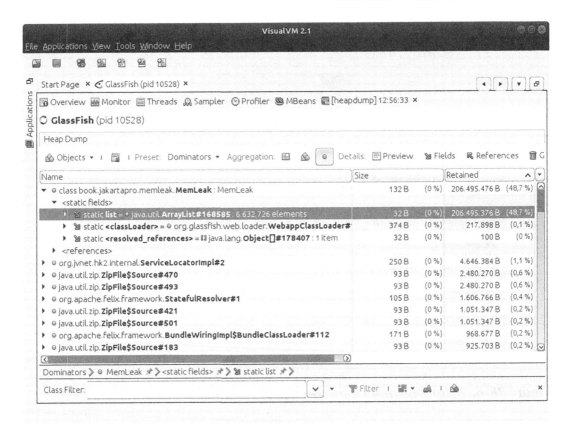

Figure 38-5. *VisualVM Dominators analysis*

Note   Expect some trial and error work. For other types of memory leaks, you'll probably have to use other settings to find something meaningful.

It is also possible to compare two heaps. The button on the left side of the Preset label allows you to select another heap dump for comparison. This is an advanced analysis technique, which can help you if looking at a single heap dump does not lead you to significant conclusions.

The Eclipse Memory Analyzer (EMA) is another quite versatile tool you can use for heap dump analysis. It is an Eclipse plugin and you can find it in the Eclipse Marketplace if you search for Memory Analyzer. After installation, there is a new perspective called Memory Analysis (choose Window ➤ Perspective). Using the toolbar, you can import a heap dump. In a wizard dialog box, check Leak Suspects Report. You will then be presented a report pointing to suspicious classes, as shown in Figure 38-6.

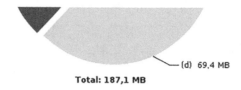

Total: 187,1 MB

▼ ⊗ **Problem Suspect 1**

The class **"book.jakartapro.memleak.MemLeak"**, loaded by
"org.glassfish.web.loader.WebappClassLoader @ 0xe77c8e88", occupies
53.058.296 (27,04 %) bytes. The memory is accumulated in one instance of
"java.lang.Object[]", loaded by **"<system class loader>"**, which occupies
53.058.184 (27,04 %) bytes.

Keywords

    book.jakartapro.memleak.MemLeak
    org.glassfish.web.loader.WebappClassLoader @ 0xe77c8e88
    java.lang.Object[]

Details »

Figure 38-6. *Eclipse memory analyzer report*

The report correctly lists the MemLeak class as a suspect. In order to investigate further, select Open Dominator Tree for Entire Heap from the dump's toolbar. The list that's presented is shown in Figure 38-7.

Class Name	Shallow Heap	▲ Retained Heap	Percentage
⌗ <Regex>	<Numeric>	<Numeric>	<Numeric>
▶ class book.jakartapro.memleak.MemLeak @ 0xe77c8fa0	8	53.058.296	27,04 %
▶ org.jvnet.hk2.internal.ServiceLocatorImpl @ 0xe1858608	136	3.729.080	1,90 %
▶ java.util.zip.ZipFile$Source @ 0xe5f15028	64	2.480.224	1,26 %
▶ java.util.zip.ZipFile$Source @ 0xe92543c0	64	2.480.224	1,26 %
▶ java.util.zip.ZipFile$Source @ 0xe5f8beb0	64	1.051.296	0,54 %
▶ java.util.zip.ZipFile$Source @ 0xe79dfd40	64	1.051.296	0,54 %
▶ org.apache.felix.framework.StatefulResolver @ 0xe065a550	64	1.022.472	0,52 %
▶ java.util.zip.ZipFile$Source @ 0xe0730630	64	925.656	0,47 %
▶ org.apache.felix.framework.BundleWiringImpl$BundleClassLoader @ 0:	104	659.256	0,34 %
▶ java.util.zip.ZipFile$Source @ 0xe57f7200	64	631.272	0,32 %
▶ java.util.zip.ZipFile$Source @ 0xe5f15730	64	631.272	0,32 %
▶ java.util.zip.ZipFile$Source @ 0xe92401e8	64	631.272	0,32 %
▶ java.util.zip.ZipFile$Source @ 0xe9255768	64	631.272	0,32 %
▶ java.util.zip.ZipFile$Source @ 0xe7270b10	64	551.624	0,28 %
▶ java.util.zip.ZipFile$Source @ 0xe8345f88	64	551.624	0,28 %
Σ Total: 15 of 160.012 entries; 159.997 more			

Figure 38-7. *EMA Dominator tree*

As was the case for the VisualVM analysis, be sure to watch out for the *retained* heap sizes. You would continue this analysis by clicking, double-clicking, or right-clicking the class in question and performing various calculations and selections. It is almost impossible to describe all the functionalities here, but the help included with the plugin should give you a starting point.

For the artificial memory leak in this chapter, you can for example right-mouse-click the MemLeak class, and then choose Leak Identification ➤ Component Report. By clicking the Retained Set label, you'll see that there is a huge number of MemLeak$A objects statically associated with this class.

CHAPTER 39

Custom Log4j Appender

Log4j appenders are responsible for writing out or sending log events to some external consumer. There are many built-in appenders you can use, so if you have special requirements, you should first check whether Log4j can fulfill your needs. Table 39-1 lists all the appenders that ship with Log4j.

> **Note** Log4j heavily changed from version 1.x to 2.x. That is why versions 2.x are commonly referred to as Log4j2. Although Log4j 1.x is still in use, I mean Log4j2 when I refer to Log4j in this chapter.

Table 39-1. *Log4j Appenders*

Name	Description
Async	Wraps another appender and sends log events asynchronously in a new thread.
Cassandra	Sends log events to an Apache Cassandra database.
Console	Sends log events to the console. Note that most Jakarta EE servers redirect the console and write console messages to the server log file instead.
Failover	Wraps a list of appenders and sends log events in order until one succeeds.
File	Sends log events to a file.
Flume	Sends log events to Apache Flume (a system for routing and aggregating large amounts of log data).
JDBC	Sends log events to a database accessible via JDBC.
JMS	Sends log events to a messaging system accessible via JMS.
JPA	Sends log events to a database via Persistence API (JPA 2.1).

(continued)

© Peter Späth 2023
P. Späth, *Pro Jakarta EE 10*, https://doi.org/10.1007/978-1-4842-8214-4_39

Table 39-1. (*continued*)

Name	Description
HTTP	Sends log events via HTTP.
Kafka	Sends log events to an Apache Kafka instance (a distributed event streaming platform).
Memory Mapped File	Sends log events to a memory mapped file. The OS takes care of writing the data to a memory region and only later syncs the data with a real file.
NoSQL	Sends data to a NoSQL database through a lightweight API. An implementation for MongoDB and CouchDB is provided.
NoSQL for MongoDB 3	Sends log events to a MongoDB v3.
NoSQL for MongoDB 4	Sends log events to a MongoDB v4.
NoSQL for CouchDB	Sends log events to a CouchDB.
Random Access File	Same as file appender, but always buffered.
Rewrite	Allows log events to be manipulated before they are sent to another appender.
Rolling File	Same as file appender, but allows for a rolling file policy.
Rolling Random Access File	Same as random access file appender, but allows for a rolling file policy.
Routing	Allows for routing events to subordinate appenders.
SMTP	Allows for sending log events as emails.
ScriptAppender Selector	Selects appenders based on a script outcome.
Socket	Writes log events to TCP or UDP sockets.
Syslog	Same as socket appender, but uses the BSD Syslog or the RFC 5424 format.
ZeroMQ/JeroMQ	Sends log events to a ZeroMQ networking system.

For more details, see the documentation at `https://logging.apache.org/log4j/2.x/manual/appenders.html`.

If none of the built-in appenders suits your needs, it is not too complicated to build a custom appender. The following sections show you how to do that.

Including Log4j

In order to use Log4j and write a custom appender, add the following dependency:

```
// Gradle
implementation 'org.apache.logging.log4j:log4j-core:2.17.2'

<!-- Maven -->
<dependency>
    <groupId>org.apache.logging.log4j</groupId>
    <artifactId>log4j-core</artifactId>
    <version>2.17.2</version>
</dependency>
```

A Statistics Appender

Consider an appender that provides statistical data about incoming log events on a daily basis. You would use this only as an auxiliary appender, which is not a problem because you can associate more than one appender to a logger. The statistics are written out just after midnight each day, when the first new message arrives. The appender class writes the following:

```
package book.jakartapro.restdate.log4j;

import java.time.LocalDate;
import java.util.HashMap;
import java.util.Map;

import org.apache.logging.log4j.core.Appender;
import org.apache.logging.log4j.core.Core;
import org.apache.logging.log4j.core.Filter;
import org.apache.logging.log4j.core.LogEvent;
```

```java
import org.apache.logging.log4j.core.appender.
    AbstractAppender;
import org.apache.logging.log4j.core.config.Property;
import org.apache.logging.log4j.core.config.plugins.*:
import org.apache.logging.log4j.spi.StandardLevel;

@Plugin(name = "StatisticsAppender",
  category = Core.CATEGORY_NAME,
  elementType = Appender.ELEMENT_TYPE)
public class StatisticsAppender extends AbstractAppender {
    public static class Stat {
        public int traceCnt = 0;
        public int debugCnt = 0;
        public int infoCnt = 0;
        public int warnCnt = 0;
        public int errorCnt = 0;
        public int allCnt() {
            return traceCnt + debugCnt + infoCnt +
                warnCnt + errorCnt;
        }
    }
    private Map<LocalDate, Stat> hist = new HashMap<>();

    protected StatisticsAppender(String name,
            Filter filter) {
        super(name, filter, null, true,
            Property.EMPTY_ARRAY);
    }

    @PluginFactory
    public static StatisticsAppender createAppender(
            @PluginAttribute("name") String name,
            @PluginElement("Filter") Filter filter) {
        return new StatisticsAppender(name, filter);
    }

    @Override
```

```java
public void append(LogEvent event) {
    updateStat(event);
    maybeWriteOut();
}

private void maybeWriteOut() {
    // If we have just one entry for today, it means
    // the last day just closed and can be written out
    LocalDate tm = LocalDate.now();
    // For testing and debugging, comment out the
    // following if()
    if(hist.get(tm).allCnt() == 1) {
        LocalDate tmx = tm.minusDays(1);
        // tmx = tm; // DEBUGGING ONLY
        if(hist.containsKey(tmx)) {
            Stat s = hist.get(tmx);
            System.out.println("STATISTICS - " +
                    tmx.toString() + "\n" +
                "all: " + s.allCnt() + "\n" +
                "trace: " + s.traceCnt + "\n" +
                "debug: " + s.debugCnt + "\n" +
                "info: " + s.infoCnt + "\n" +
                "warn: " + s.warnCnt + "\n" +
                "error: " + s.errorCnt + "\n");
        }
        // Add some cleanup code here, removing old
        // entries from hist...
    }
}

private void updateStat(LogEvent event) {
    LocalDate tm = LocalDate.now();
    Stat stat = hist.computeIfAbsent(tm,
        tm1 -> new Stat());
    int l = event.getLevel().intLevel();
    if(l == StandardLevel.TRACE.intLevel())
```

```java
            stat.traceCnt++;
        else if(l == StandardLevel.DEBUG.intLevel())
            stat.debugCnt++;
        else if(l == StandardLevel.INFO.intLevel())
            stat.infoCnt++;
        else if(l == StandardLevel.WARN.intLevel())
            stat.warnCnt++;
        else if(l == StandardLevel.ERROR.intLevel())
            stat.errorCnt++;
        else if(l == StandardLevel.FATAL.intLevel())
            stat.errorCnt++;
    }
}
```

The @Plugin annotation is the central configuration element. It qualifies the class to use as an appender. This example uses the abstract class AbstractAppender. This helps avoid writing boilerplate code. The most important method is append()—this is where new log events arrive. As usual, you can find more information about the annotations and the base class in the API documentation.

A basic Log4j2 configuration file using the XML format would read as follows:

```xml
<?xml version="1.0" encoding="UTF-8"?>
<!-- Note the "packages" attribute - it must match -->
<!-- the plugin class' package name              -->
<Configuration xmlns:xi="http://www.w3.org/2001/XInclude"
    packages="book.jakartapro.restdate.log4j"
    status="DEBUG">
  <Appenders>
    <Console name="ConsoleAppender"
        target="SYSTEM_OUT">
      <PatternLayout
          pattern="%d [%p] %c{1} - %m%n"/>
    </Console>
    <StatisticsAppender name="StatisticsAppender"/>
  </Appenders>
  <Loggers>
```

```
    <Root level="DEBUG">
        <AppenderRef ref="ConsoleAppender" />
        <AppenderRef ref="StatisticsAppender" />
    </Root>
  </Loggers>
</Configuration>
```

Name it log4j2.xml and, for a web application built via Maven or Gradle, save it in the src.main.resources. folder Finally, add Log4j logging instructions to any application to test the appender.

Index

© Peter Späth 2023
P. Späth, *Pro Jakarta EE 10*, https://doi.org/10.1007/978-1-4842-8214-4

Printed in the United States
by Baker & Taylor Publisher Services